Biological Control in Plant Protection

A Color Handbook

Second Edition

Biological Control in Plant Protection

A Color Handbook

Second Edition

Neil Helyer
Fargro Ltd, West Sussex, UK

Nigel D. Cattlin
Holt Studios International Ltd

Kevin C. Brown
Ecological Consultant, Plymouth, UK

CRC Press
Taylor & Francis Group
Boca Raton London New York

CRC Press is an imprint of the
Taylor & Francis Group, an **informa** business

Consultant editor: Dr. John Fletcher

CRC Press
Taylor & Francis Group
6000 Broken Sound Parkway NW, Suite 300
Boca Raton, FL 33487-2742

© 2014 by Taylor & Francis Group, LLC
CRC Press is an imprint of Taylor & Francis Group, an Informa business

No claim to original U.S. Government works

Printed on acid-free paper
Version Date: 20140108

International Standard Book Number-13: 978-1-84076-117-7 (Hardback)

Library of Congress Cataloging-in-Publication Data

Helyer, Neil.
   Biological control in plant protection : a color handbook / Neil Helyer, Nigel D. Cattlin, and Kevin C. Brown. -- 2nd edition.
     pages cm
   Originally published: A color handbook of biological control in plant protection. Portland, Or. : Timber Press, copyright 2003.
   Includes bibliographical references and index.
   ISBN 978-1-84076-117-7
   1. Agricultural pests--Biological control--Handbooks, manuals, etc. 2. Biological pest control agents--Handbooks, manuals, etc. 3. Insects as biological pest control agents--Handbooks, manuals, etc. 4. Natural pesticides--Handbooks, manuals, etc. I. Cattlin, Nigel D. II. Brown, Kevin, 1959 August 9- III. Title.

SB975.H46 2014
363.7'8--dc23
                                          2013035155

Visit the Taylor & Francis Web site at
http://www.taylorandfrancis.com

and the CRC Press Web site at
http://www.crcpress.com

Contents

CHAPTER 5
Beneficial pathogens

CHAPTER 6
Biological control in perspective

Preface

Since the first edition of this book there have been dramatic increases in the commercial use of integrated crop/pest management (ICM/IPM) methods for pest and disease control on a wide range of crops throughout the world. This has resulted in an enormous demand for mass-reared biological control agents, and the knowledge and technology to do this has improved with the demand. Some producers concentrate on a narrow range of organisms reared under highly specific conditions. Others offer a broader range. The development of ICM/IPM compatible pesticides that frequently control specific pests while having minimal side effects on beneficials allows a greater range of biological control agents to be used over a longer period and with greater success. The majority of protected crops (edible and ornamental) are now produced using these methods to control their major pests and diseases. Biological agents for the control of diseases are being used more frequently, particularly on high-value edible crops like strawberry. Similarly, broad-spectrum pesticides are being used less extensively on field and orchard crops; consequently, growers and advisors are now finding a greater frequency and range of naturally occurring beneficials.

All these factors come together under the ICM/IPM banner as a more environmentally friendly and sustainable method for long-term pest and disease control. The majority of mass-reared beneficials can be used by amateur growers and are usually available by mail order for direct delivery and use in the garden, conservatory, or greenhouse. The reduction of pesticide applications in the home garden allows more natural enemies of common pests to establish. This book should help both amateur and professional growers identify many of these organisms.

Biocontrol is about the control of plants (weeds), animals (insect and mite pests but could include slugs and nematodes pests), or diseases by using another living organism, be it herbivore, predator, parasitoid, parasite, or microorganism. Unlike pesticides, biocontrol seldom kills all of the target species but aims to manage them to a level that is below the economic damage threshold.

Classical biocontrol is where a beneficial organism is released on a one-off basis—usually to control a new pest that has been introduced from elsewhere, thus reestablishing it with its former natural enemies. Frequent or inundative release of biological control agents is used in glasshouses or annual crops where the beneficial may reproduce over a period of time but dies out at the end of the year. In some cases, particularly when microorganisms are used, they are treated like a pesticide with little residual action or subsequent reproduction. Conservation biocontrol aims to change our modern agricultural practices of a single monoculture crop grown in bare soil to one of supporting native beneficial insects by using additional plants to provide shelter and alternative sources of food.

A prerequisite for classical and inundative biocontrol is that the beneficial organism has to be available commercially or financed through some government-sponsored national release scheme. It must also be specific against the pest to be controlled. This book covers many commercially produced beneficials; because of legislation within each country, certain beneficials may or may not be permitted for use. Biopesticides is a term frequently used when a living organism is applied in similar ways to chemical or conventional pesticides. Several pathogens (disease-causing organisms) are now available to control and help prevent pest outbreaks on a range of crops. These can include bacteria, fungi, and viruses. The majority occur naturally, but when they are used as a pesticide, similar registration conditions to chemical pesticides apply. This has been seen as restriction to their development. However, since over 60% of the active ingredients of pesticides have been withdrawn, the new biological controls are now regarded as a safer and a more sustainable alternative.

A word of warning: this book illustrates and gives information about biological control agents that may be unlicensed species in some countries. Some of these can be obtained on the Internet or from other sources. Purchasers of these should be aware that if unlicensed biocontrol agents are used, the purchaser is liable to a fine (currently £1,000 per offense in the UK).

This book was first published in 2003 and was twice reprinted. The second edition has been completely revised with a number of new photographs, additional pest and beneficial organisms, a new section on the practical aspects and application of biological control, and a final chapter that puts biological control in perspective. The original idea for the book developed from a suggestion by Dr. Paul Jepson, now at Oregon State University. The objective was to produce a handbook containing profiles and color photographs of as many examples of biological control organisms from as wide a global area as possible. We have continued with this theme in the second edition and have added further information to help the reader more fully understand the concepts and practice of biological control and integrated pest and disease management.

Descriptions of the biocontrol organisms are divided into four sections: species characteristics, including organism size, host food, and closely related species; life cycle; crop/pest associations; and influences of growing practices. The section on crop/pest associations describes how and when the organism attacks its prey, the crops and environments in which it is likely to be found, and whether it is commercially available. The section on the influence of growing practices completes each profile by summarizing how growers can make best use of these natural enemies, and often makes mention of harmful, safe, and IPM-compatible pesticides.

Although all the organisms occur naturally in various parts of the world and several are commercially mass produced, many can only be found in their natural environment and usually close to their food sources. We therefore thought that a short section on the pests was essential, since all the natural enemies require a host for their survival.

This handbook will be useful to advisors, extension officers, educators and research workers, and to all growers, whether amateur or professional, with an eye for the environment, no matter how large or small the area under production. Finally, we would like to make specific acknowledgments to Dr. Mike Copland and Dr. John Fletcher. Both have read the entire book and made many useful suggestions, Mike has also written the final chapter, which we feel improves the whole book.

NEIL HELYER

NIGEL CATTLIN

KEVIN BROWN

Acknowledgments

We wish to thank the following for providing technical information and organisms to photograph:

BASF Agricultural Specialties, Harwood Rd., Littlehampton, UK

Dr. Ian Bedford, John Innes Centre, Norwich, UK

Jude Bennison, ADAS, Boxworth, Cambridgeshire, UK

Dr. Dave Chandler, School of Life Sciences, University of Warwick, UK

Entocare, Wageningen, the Netherlands

Dr. David R. Gillespie, section head, IPM and research scientist, Agriculture and Agri-Food Canada

Dr. Richard GreatRex, Syngenta Bioline, Little Clacton, Essex, UK

Dr. Andrew Halstead, senior entomologist, Royal Horticultural Society Gardens, Wisley, UK

Dr. Paul Jarrett, School of Life Sciences, University of Warwick, UK

Professor Paul Jepson, Department of Entomology, Oregon State University, Corvallis, Oregon

Dr. Garry Keane, School of Life Sciences, University of Warwick, UK

Mike Mead-Briggs, Mambo-Tox Ltd., Southampton, UK

Dr. Graeme Murphy, Ontario Ministry of Agriculture and Food, Vineland, Ontario, Canada

Dr. Manuele Ricci, Biologics, Bayer Crop Science, Deruta (PG), Italy

Dr. Richard Shaw, CABI Europe-UK Centre, Bakeham Lane, Egham, Surrey, UK

Simon Springate, NRI, University of Greenwich, UK

Dr. Marylin Steiner, Biocontrol Solutions, Mangrove Mountain, New South Wales, Australia

Suzanne Stickles, Wyebugs, Kent UK

Les Wardlow, Ruckinge, Kent, UK

About the authors

Neil Helyer joined the Glasshouse Crops Research Institute as an entomologist in 1976, working on several projects including developing mass rearing techniques for pest and beneficial organisms, evaluating efficacy of biological control agents, screening pesticides for side effects on biological agents, and screening pesticides and biological control organisms against Western flower thrips. He joined Fargro Ltd. as their IPM specialist in 1995, when the research station transferred to Wellesbourne and became Horticulture Research International. Currently, Neil visits horticulturalists to develop and monitor IPM programs, the majority of which are specifically designed for each site and crop. These include protected salad crops (cucumber, herbs, sweet pepper, and tomato), soft fruit, ornamentals (cut flower and pot plants), and hardy nursery stock, as well as botanic gardens (RBG Kew, RHS Wisley, etc.) and interior plant landscapes (atriums). In line with the audit requirements of some producers/suppliers, Neil holds a BASIS certificate and ensures that his technical information base is kept up to date by being a member of the BASIS Professional Register.

Nigel Cattlin joined ICI's Plant Protection Division at Jealott's Hill Research Station in 1963. He later formed and led its photographic unit, carrying out research projects with high-speed, time-lapse, and aerial photography. He also photographed a wide range of weed plant species, pest organisms and the damage they cause, and disease symptoms in crops. These were used to aid identification in technical literature and grower information and for advertising and marketing new agrochemical products. In 1981 Nigel established Holt Studios, an independent company providing a specialist photographic service to the agricultural industry, government orga-

nizations, and publishers. He built up an extensive collection of photographs ranging from photomicrographs to landscapes to illustrate literature covering international agriculture, horticulture, and gardening. This resource is widely used for text and reference books, trade and consumer magazines, and advertising. Nigel's photographs are now with many leading international image libraries. With Holt Studios he also undertook commissions for several multinational chemical companies, photographing pests, disease and weed species, and their effects on many of the world's major crops. In the early 1990s he collaborated with Professor Jepson, then at Southampton University, in the early work for the first edition of *Biological Control in Crop Protection* and has been coauthor of three other books in the Manson Color Atlas series. Holt Studios' main business was transferred to another agency in 2005, but Nigel continues to take photographs in his specialized fields of interest.

Dr. Kevin Brown joined the Sittingbourne Laboratory of Shell Research Ltd. in 1986 as an ecotoxicologist working on the effects of insecticides on beneficial and nontarget arthropods. He designed and conducted field studies in a range of tree and broad-acre crop systems in Northern and Southern Europe. He was a founding member of the Beneficial Arthropod Testing Group (BART) in 1988 and is an author of many of the current regulatory ring-tested methods for nontarget arthropods. In 1989 he established Ecotox Limited, a contract testing facility based in Devon to conduct laboratory tests and field trials on a commercial basis. Although he originally specialized in large-scale field trials with arthropods, his interests have broadened to include aquatic organisms, earthworms, birds, and mammals. After a brief spell with a major consultancy firm, he now works as an independent ecotoxicologist and environmental consultant conducting and refining risk assessments, monitoring higher tier studies on behalf of clients, and participating in multifacility research projects. He is a member of the UK Environmental Panel and has a particular fondness for staphylinid beetles.

The practice and application of biological control

INTRODUCTION

Biological control agents are also known as natural enemies or beneficial organisms. This implies they are naturally occurring, often in the locality of the pest and of benefit to growers in terms of pest control. There are numerous definitions of what a pest is; the majority agree that it is an organism, usually numerous, that is unwanted and has negative impact on the growth of plants. A pest of plants can therefore be an injurious insect, mite, mollusk, rodent, or any other of the myriad organisms that feed on or damage plants. In addition to pests, there are many pathogens including fungal, bacterial, phytoplasmal, and viral organisms.

The main pest organisms infesting plants include aphids, caterpillar, sawfly, spider mite, thrips, whitefly, and various beetles. These common pests form the basic food chain for many beneficial organisms with several species from different genera and orders feeding on one or another of these plant-feeding pests. Moth eggs, for example,

can be attacked by the minute *Trichogramma* spp. wasp, changing them to a much darker color when the parasitoid pupates inside the moth eggs. Caterpillars may be attacked by numerous parasitoid wasps or eaten by predators. Chapter 3 of this book details many of the common plant pests and indicates their natural enemies, while Chapter 4 covers many arthropod natural enemies and Chapter 5 several of the microbial beneficials. Specific microbial pathogens often provide the greatest degree of control by infecting whole populations and spreading as an epizootic disease. Minor pests such as leaf and plant hoppers, leaf miner, mealybug, and scale insects can be equally damaging but tend to have more specialized beneficials that feed only on a particular pest genus or even species.

Once the pest and biological control organism has been found and identified, better utilization and improved pest control can be established. All too frequently, a broad-spectrum pesticide will be sprayed that should kill the pest, but will undoubtedly kill the natural enemies as well. Several broad-spectrum, synthetic pyrethroid insecticides can remain active against a pest species for 3 to 4 weeks, but can kill beneficials for up to 12 weeks, drastically reducing the number of natural enemies that can survive to help reduce further pest outbreaks.

There are three major categories of biological control organisms: parasites and parasitoids, predators, and pathogens.

Parasites and Parasitoids

A parasite is an organism that forms a nonmutual relationship between organisms of different species; the parasite benefits at the expense of the other, which is known as the host. The parasite is dependent on its host for part or most of its development and survival. There are various types of parasitic relationships. Obligate parasites cannot survive independently of their hosts (rusts and powdery mildews), whereas facultative parasites (gray mould and leaf spot diseases) can. Parasites may or may not kill their host (e.g., mistletoe is a parasitic plant that requires a living host plant for its survival). Microparasites including bacteria and viruses can infect a host, causing a disease (e.g., tulip-break or mosaic virus, which produces dramatic color streaks in the flowers). Most common plant diseases such as powdery mildews, leaf spots, and rusts are deleterious to the survival of the host plant.

Parasitoids usually have a free living adult stage and a larval stage that develops on or within a single host organism, ultimately killing the host, which is a great benefit in terms of biological control. Parasitoids can be further classed as ectoparasitoids that live and attack from the outside of the host and endoparasitoids that develop from an egg laid within the host. Examples are, respectively, *Eretmocerus eremicus* and *Encarsia formosa*, which are both parasitoids of whitefly larvae. The adults of many parasitoid wasps can also kill and feed as predators, often killing one or more insects each day and contributing greatly to their overall efficacy as biological control agents. Known as destructive host feeding, this is done by the female inserting her ovipositor (used for egg laying) several times into the same host, wounding and slicing the internal organs; she then feeds on the dying insect. Parasitoid wasps can be further divided, by studying their ovipositional lifestyles, into either idiobionts or koinobionts.

Idiobionts totally paralyze their host; the majority are ectoparasitoids that deposit an egg (or more) next to the host that develops outside the prey body, sucking nutrients from the living but paralyzed body (e.g., the leaf miner parasitoid *Diglyphus isaea*). Koinobionts are generally endoparasitoids and cause no signs of parasitism until the host has reached a suitable size and the young hatched parasitoid larva begins to grow and consume the host. The leaf miner parasitoid *Dacnusa sibirica* is a koinobiont, developing only once the leaf miner has pupated; however, the majority attack the young stages of exposed hosts. Both parasitoid and host develop together through to pupation (i.e., aphid parasitoid *Aphidius* spp.).

Several caterpillar parasitoids are known as gregarious parasitoids, whereby several individuals can develop as a single brood from the body of an individual host; *Cotesia glomerata* sometimes producing 20–40 larvae from a single host caterpillar. This also occurs in the leaf miner parasitoid *Diglyphus isaea*, which can produce two or even three individuals from a single host, making this wasp extremely useful in commercial situations. Hyperparasitoids, however, use another parasitoid as a host, usually attacking the larval stage within the primary host and killing it as the second parasitoid develops. This can result in a resurgence of pests (see *Aphidius colemani*).

Predators

These attack and feed entirely from the outside of the host—usually by piercing through the host organism's cuticle and damaging its internal organs before sucking it dry, although some larger predators devour the entire body. In many predatory insects only the larval stage actively feeds on prey; the adults survive on insect honeydew, plant nectar, and pollen. However, with predators such as ladybirds, the larval stages and adults actively feed on prey.

Pathogens

These are disease-causing organisms that kill by parasitism either directly or as a result of toxins that destroy the hosts' internal organs, allowing the pathogen to reproduce. In almost all instances the host organism is killed and further beneficials are produced. One notable exception is the commercial use of *Bacillus thuringiensis* (*Bt*), as most products contain dead bacteria and it is their protein toxins that kill the target pest with little if any establishment in the treated area.

There are several named isolates of *Bt* that have activity against a range of seriously economically damaging pests. These include various Lepidoptera (caterpillar), Coleoptera (beetles; specifically, the Colorado potato beetle, *Leptinotarsa decemlineata*), and Diptera (mosquitoes, fungus gnats, and blackflies [Simuliidae] but not Homopteran aphids). Through the techniques of genetic engineering, the *Bt* toxin gene has been transgenically transferred into various plant genomes. As the *Bt* toxin is produced throughout the plant, any target pest larva feeding on the plant will rapidly perish. The technique is considered environmentally beneficial as it replaces the widespread use of several pesticides. However, there is an argument that the presence of pest organisms feeding on plant material helps in recycling nutrients back to the soil. Chapter 5 gives details of nematodes, fungi, bacteria, and viruses as beneficial organisms.

STRATEGIES FOR BIOLOGICAL CONTROL

There are four major strategies for the use of biocontrol.

Conservation or Preservation Biological Control

This method is based on exploiting existing natural enemies by modifying the environment to encourage establishment and survival of greater numbers of beneficials. This may involve developing areas that act as refugia for beneficial organisms by providing suitable plants and sites for overwintering and early pollen production. In order to aid the survival of natural enemies, cultural practices may also need to be changed by stopping the use of certain pesticides that may be broad spectrum and have long persistence.

Creating refuges or providing suitable habitats for biological control organisms is something the cottage gardener has been doing for generations, and this technique is now being practiced on arable farms in the form of uncultivated field headlands (Figure 1.1). Also, some horticultural nurseries plant small areas with specific plants to encourage beneficials and their pest hosts that can migrate into the greenhouse. These provide an additional degree of control of the pests of the crops. Generalist predators such as lacewing larvae (*Chrysoperla carnea*) are sometimes used on hedges surrounding the holding. These predators feed on most soft-bodied prey such as aphids, young leaf hopper nymphs, moth eggs, etc. This is done to reduce the number of

Figure 1.1   Flowering *Phacelia tanacetifolia* between two fields of cereals. A wide strip of uncultivated ground creates refugia for many insects and spiders.

pests before they migrate into the production areas. It also helps to establish these extremely useful beneficials. This technique can also be used by many home gardeners (Figure 1.2). Similarly, an established garden with mixed permanently planted areas, as well as seasonally flowering plants, will allow a diverse range of wildlife to establish; this would include a few pest species but also several beneficial species (Figure 1.3). This creates a "natural" balance and rarely do pests get out of hand, causing just minimal plant damage before beneficial organisms take over to feed on the pests.

Examples of plants that encourage beneficials include Compositae or Asteraceae such as chamomile (*Matricaria chamomilla*), leopard's bane (*Doronicum* x *excelsum*), Michaelmas daisy (*Aster* spp.), and goldenrod (*Solidago virgaurea*), which are well known for their flowers later in summer and will encourage hoverflies, robber flies, and predatory wasps to remain in the garden. Several low growing herbs from the Lamicaceae family, including mint (*Mentha* spp.), rosemary (*Rosmarinus officinalis*), and thyme (*Thymus* spp.), are ideal host plants for ground beetles and parasitoid wasps. The presence of early flowering, pollen-producing plants will allow predatory mites and insects to establish in high enough numbers to prevent major pest outbreaks. Many of the Umbelliferae, such as fennel (*Foeniculum vulgare*), angelica (*Angelica archangelica*), carrot (*Daucus carota*), and parsnip (*Pastinaca sativa*), are grown for medicinal and food uses, but all provide an excellent source of nutrition for beneficial insects.

Fully planted refugia should contain a few plants from each of the preceding types; several weeds such as nettles can also be included. Stinging nettles (*Urtica dioica*) will allow the common nettle aphid (*Microlophium carnosum*), which is host specific (Figure 1.4), to establish as soon as the nettles begin to grow, and these in turn provide food for various parasitoids, pathogens, and predators. Some dead wood, ideally with loose bark, will provide a suitable overwintering site for adult ladybirds and other beetles. Loosely packed straw in a flowerpot or a small bundle

Figure 1.2   A garden with uncut borders creates an informal refugium for predators.

Figure 1.3   A formal English garden with permanently planted areas and seasonally flowering plants; ideal for diverse wildlife establishment.

Figure 1.4   Nettle aphids (*Microlophium carnosum*) on a nettle leaf (*Urtica dioica*) provide alternative food sources for predators and parasites; lower left: young hoverfly larva.

Figure 1.5 A simply constructed overwintering shelter for garden predators using old bamboo canes.

of bamboo canes or corrugated cardboard hung in a dry shed or garage will provide a similar overwintering refuge for adult lacewings and ladybirds (Figure 1.5). Large colonies of aphids, which can appear during the spring months, produce sticky honeydew laden with kairomones that attract and feed many parasitoids and predators. Biological control organisms may then be able to locate the pest. Subsequent generations of beneficial organisms may then provide sustained pest control for the whole season.

Augmentation biological control involves the periodical release of beneficial organisms to control a pest population, frequently using commercially reared agents. There are approximately 230 different species of beneficial arthropods in commercial production throughout the world (van Lenteren 2011); some are released in very large numbers to gain a rapid pest control effect.

Importation/Classical Biological Control

Beneficials are introduced to an area, country, or continent to control a pest that does not have effective, native, natural enemies. The pest may have been imported on plants or other products from almost any country or state and consequently requires beneficials from that locality. Once introduced, there is an expectation of long-term or even permanent establishment to maintain the equilibrium.

Inoculative Control

The aim is to establish the biocontrol agent on a short-term basis. Such agents may be native or introduced under license or permit. The biocontrols may reproduce and persist while the pest is present, but not form a permanent relationship. This method is used more extensively in greenhouse and interior settings than in large-scale open field crops. For instance, the periodic (weekly or fortnightly) releases of the parasitoid *E. formosa* to whitefly-susceptible plants is done to protect many greenhouse crops. On the other hand, the predatory mite *Phytoseiulus persimilis* tends to be used more as a curative treatment following the first signs of spider mite damage on plants. Many commercial growers have a biological control schedule devised that accommodates the seasonal changes in plant development, pest likelihood, and plant susceptibility.

Inundative Control

With this method the pest is overwhelmed by mass applications of commercially reared beneficials but with no expectation of long-term control of the pest's progeny. This form of biological control is also referred to as "biopesticide applications" as the majority of these agents are applied using slightly modified conventional pesticide application equipment. Microbial pathogens are applied in this way, often using vast populations that are often produced using fermentation type processes (Figure 1.6). For instance, entomopathogenic

Figure 1.6   Large-scale fermenters for nematode production. (Photograph courtesy of BASF Agricultural Specialties, Littlehampton, W. Sussex, UK.)

nematodes are applied at rates of 0.5 to 1 million per square meter to control many soil pests; some species and formulations can also be sprayed to the foliage as a living insecticide. The minute parasitoid wasp *Trichogramma* spp. can be introduced at rates of several million per hectare to control a range of moth eggs. To introduce the numbers required over a short period, the parasitoids can be produced on artificial eggs, stored for several months with little loss of viability, and released in biodegradable packs distributed by light aircraft.

INTEGRATED CROP/ PEST MANAGEMENT

Biological control, either alone or as part of an integrated crop/pest management (ICM, IPM) program, is continually developing as new organisms and compatible pesticides are found and commercialized. These, together with cultural and physical control methods, are integrated to produce the best possible pest- and disease-control strategy. There are several factors driving this evolving science, including the arrival of new pest organisms that may originate from other countries, without native natural enemies to control them. Also, existing pest populations can gain tolerance or resistance to chemical pesticides, resulting in reduced levels of pest control. New local or national legislation can be a factor by resulting in the withdrawal of pesticides, leaving growers little option but to attempt some form of IPM and biological control.

Many marketing organizations, in particular supermarket chains, their suppliers, and crop protection specialists, collaborate in global G.A.P (global good agricultural practice, www.globalgap.org). This is a nongovernmental organization that sets voluntary standards for the certification of agricultural products around the globe. The products include many food items (agricultural and horticultural), livestock, and flowers. One of the guidelines for growers/suppliers concerns IPM and biological control in conjunction with nonchemical control. Regular audits are done to ensure these criteria are met; some are more restrictive about the use of pesticides than national and international regulations. However, some pesticides are acceptable and can be used in certain circumstances. Site-specific pesticides such as pymetrozine, which affects the feeding activity of sap-sucking insects including aphids, leaf hopper, and whitefly, is safe to the majority of biocontrol agents. Harmful pesticides may also be used if applications are separated by space and time (i.e., treating just the section of the plant where a pest is most active to reduce pest numbers to below damaging levels, allowing beneficials to catch up). This usually involves spraying the head or top of plants where the pest accumulates in highest numbers. This situation applies with adult whiteflies, which are found mainly on the newest growth where they lay their eggs. The parasitoid *Encarsia formosa* cannot attack adults, requiring the third larval instar for successful parasitism.

Thus, although difficult, it would be perfectly safe to treat the heads of plants with a pesticide to reduce the number of adult whitefly while having minimal effect on parasitoid activity lower in the crop canopy. Short persistence products, such as natural pyrethrum, soap, and plant-derived pesticides, that are generally harmful to the majority of biocontrol organisms can usually be integrated in this way. Similarly, the pupal or "mummy" stage of parasitized aphids is well protected against most of these short persistence pesticides.

However, a single application of a site-specific product is usually enough to reduce a pest outbreak and allow the beneficials to regain control, whereas multiple applications of many naturally derived organic pesticides such as pyrethrum may be required as they will kill a wider range of natural enemies. In many organic production units, beneficials are introduced to augment the existing population of natural enemies. However, the rate of beneficials introduced is higher than comparable areas grown using IPM techniques and selective conventional pesticides.

PRACTICAL ASPECTS OF BIOLOGICAL CONTROL

Arthropod biocontrol organisms have been found by several means, but all involve good observation among a known pest population. Often, serendipity or observations at the "right place, right time" have been the main source of currently mass-reared beneficials. However, pathogenic biocontrol organisms are actively searched for in order to find the most active strain. Different strains of the same organism may be marketed under different trade names.

Before a potential biocontrol agent becomes a commercial product, it must be evaluated for efficacy against the target pest and nontarget organisms. This is particularly important for non-native organisms that may become a pest if they escape from the treated area. The next step is the successful and economic rearing of the organism, which includes collecting or harvesting, storage, packaging, and transport so that the biocontrol agent arrives in a healthy, viable condition. Many predators and most parasitoids are reared on their normal host organism; this frequently involves producing clean, healthy plants and infesting them

with a pest. When the pest population has increased to the desired numbers, the beneficial organism is introduced, often in very high numbers over a short period of time to produce a synchronous population. These are allowed to develop to a predetermined stage before the plants are processed to remove the beneficial (these processes are often well guarded secrets) without causing undue harm to the beneficial. After any remaining pest organisms are removed, they may be stored under precisely managed conditions before they are packed for transport and distribution by the user.

The whitefly parasitoid *Encarsia formosa* is usually reared on the glasshouse whitefly *Trialeurodes vaporariorum*, which in turn is reared on tobacco plants. When they were first commercialized in the 1930s, whole leaves of parasitized black pupal scales were wrapped between sheets of tissue paper and posted to users. Later, the leaves were cut in small (2 cm) sections that were stuck to a card for hanging on plants. Currently, almost all producers stick pure black pupal "scales" removed from leaves onto strips of card (Figure 1.7) and perforated for ease of distribution, with a cut on one side for hanging onto plants. Aphid parasitoid wasps are traditionally produced on plants infested with aphids and harvested as the pupal "mummy" stage. Several parasitoid wasps are now produced

Figure 1.7 Parasitoid wasps (*Encarsia formosa*) emerging from a distribution card used for whitefly control.

in a plant-less system, which lends itself to factory production methods. The parasitoids are packaged for distribution as either pure mummies or as emergent adults with an inert carrier material such as sawdust or vermiculite. Predatory insects such as *Aphidoletes aphidimyza* pupate away from the leaf on which they developed. They can be collected as mature larvae over shallow trays of fine sand in which they form a pupal cocoon; these are placed in a moist carrier material for distribution within a crop (Figure 1.8). Lacewing larvae (*Chrysoperla carnea*) are frequently distributed in a buckwheat husk carrier material that forms half-moon shells in which the predatory and cannibalistic larvae hide.

Several predatory mites (not *Phytoseiulus persimilis*) are produced on a factitious host, usually a mite that feeds on bran or dried fruit and that is not a plant pest. By this method vast numbers can be produced in a relatively small space under almost laboratory conditions, and they can be packaged in tubes or bags containing the factitious mite and vermiculite. The predatory mite *P. persimilis* is reared on its natural host spider mite, usually *Tetranychus urticae,* the two spotted spider mite and packaged in vials (Figure 1.9) or bottles containing either vermiculite or wood shavings as an inert carrier that improves distribution over plants.

Many of these mites can be packaged in small sachets that contain a complete breeding population with the factitious host, its food source, and the predatory mites. These are known as continuous rearing system or controlled release system (CRS) sachets as they can produce predatory mites continuously over a 6- to 8-week period to inoculate and protect plants against various pests. The predatory mite *Amblyseius cucumeris*, which is used against thrips, can be distributed in several methods, including, loose material for hand or machine application, sachet applications as single units, twin units, and some in extremely long lines of up to 160 m long for use on bed-grown crops. The various methods of introduction have been developed to meet the needs of various cropping situations. For example, when used on cucumbers, the single sachet of *A. cucumeris* could take up to 10 hours per hectare to apply; the development of the twin (inverted "V") sachets (Figure 1.10) more than halved the time taken to treat the same area. Machine-applied *A. cucumeris* similarly decreased the time taken when compared to hand application and also made the distribution more even over the treated area (Figure 1.11). The long strips of *A. cucumeris* have been developed for machine application using a conventional gantry sprayer as used for pesticide treatments; this method

Figure 1.8   Blister pack containing *Aphidoletes aphidimyza* cocoons in moist vermiculite carrier; note adults inside transparent blister and open back from where they emerge.

Figure 1.9   Vial of *Phytoseiulus persimilis* being released with a vermiculite carrier.

Figure 1.10 Inverted "V" sachet containing predatory *Amblyseid* mites placed on a conservatory plant, *Sparmannia africana*.

Figure 1.11 Mechanical application of predators, with a handheld blower.

allows whole bays in a greenhouse to be treated at once (Figure 1.12). The CRS can thus protect a crop of chrysanthemums (and many other crops) for the majority of their growing period by making just a single application. *A. cucumeris* can also feed on several other food sources, such as glandular leaf

Figure 1.12 Mechanical application of predatory mites in long lengths of controlled release system (CRS) paper sachets by a modified overhead gantry sprayer.

hairs, pollen, and nectar; they can therefore effectively be introduced to a crop before the pest occurs and frequently provide excellent season-long protection from thrips attack.

Parasitoid wasps of aphids, caterpillars, whiteflies, etc., require the presence of the host to survive and reproduce. However, they are frequently introduced to commercial crops before the pest can be detected on plants and are thus used as a measure to prevent establishment of emerging or migrating pests. As the pest numbers increase due to migration from other plants or outside, the beneficials are already present and high levels of control are achieved. Frequently, fresh introductions of these parasitoids are made each week to maintain the control with an increase in rates made to areas of higher pest levels. This method of applying biological control agents usually requires a schedule of introductions over a period of weeks and some degree of monitoring or scouting to locate any pest "hotspots" and determine its efficacy. IPM-compatible pesticides or increased numbers of beneficials can be applied to treat any increases in pest levels. Many smaller scale producers and amateur growers wait until the pest is present before introducing the beneficials, frequently applying a short persistence or physically active pesticide

Figure 1.14   Taking a photograph with a mobile phone looking through a "linen tester" type of hand lens.

Figure 1.13   Delta trap with sticky base and pheromone lure to trap male pea moths (*Cydia nigricana*).

prior to initiating a program of biocontrol introductions.

PEST MONITORING

Monitoring plants should be done by visual assessment of leaves, stems, and, if necessary, roots. Sticky traps are available in a range of colors to catch various flying insects. They work by the light reflected from the surface, which attracts adult flying insects to the sticky surface. For winged aphids, leaf hoppers, thrips, and white-flies, the traps should be hung vertically but for adult flea beetles, leaf miners, scatella and sciarid flies, horizontal traps that are sticky side up are best. Pheromone lures (Figure 1.13) are available for a wide range of adult insects. Most of these use a sex attractant to catch males and are specific to

each species, so correct pest identification is critical in order to choose the correct lure. Specific lures are available for numerous moth species, citrus mealybug, Western flower thrips, various midge species, and a few other major pest organisms.

A hand lens or magnifying glass is one of the most important tools a grower can have and 8×–10× power magnification is ideal. Linen tester type lenses are convenient as they can be folded away and prefocused from the lens to the target area; as such, they may be placed directly over a target that can be viewed through the lens. The majority of mobile/cell phones can be used to take a photograph through such a lens (Figure 1.14) that can be e-mailed and kept on a computer to build a library of useful images. Close inspection of pest colonies can often reveal the eggs of a predator or the first signs of parasitoid activity. Further inspection over a period of days will invariably show biological control in action.

Biological control in various cropping systems

INTRODUCTION

The number of pests reported on crops is increasing throughout the world; this may be due to changes in environmental conditions, different cropping patterns, plant movement, or a combination of several factors. In many instances the natural enemies of these pests lag behind. Frequently, where a suitable biological control agent is known, governmental regulations can severely restrict its use. Beneficial organisms are now being released under license schemes whereby the user must hold a copy and abide by its strict wording. Licenses are issued by the country or state wishing to introduce a nonindigenous biocontrol agent, usually following risk assessments to evaluate its potential environmental impact.

The use of microorganisms (bacteria, fungi, viruses, etc.) as biological control agents is increasing due to the feasibility of their mass production, pest specificity, and relative ease of use (the majority are applied as a spray similar to pesticides). There is a worldwide trend to reduce the use of conventional,

synthetic pesticides to such an extent that there is now serious concern for future crop production. Many older pesticides have been withdrawn and more recent ones, such as the neonicotinoid products, are under scrutiny due to their long persistence and potential food chain activity on nontarget organisms. The European Parliament commissioned a report, "The Consequences of the 'Cut Off' Criteria for Pesticides: Alternative Methods of Cultivation" (Chandler 2008), that stressed the requirement for placing integrated pest management (IPM) at the center of crop protection policy. Since 2014 all commercial agricultural and horticultural production throughout the EU must include an element of IPM as a method of pest and disease management.

IPM brings together all methods of pest and disease control, including cultural techniques, monitoring, the use of resistant cultivars (natural selection or genetic modification where allowed), and the regulated use of pesticides. Biological control becomes the next line of defense, either by introducing beneficial organisms or by allowing natural enemies to establish, often as a consequence of the reduction in use of broad-spectrum pesticides. As a general guide, the majority of fungicides are safer to beneficial insects and mites; however, they may have serious side effects to fungal pathogens used to control insects, mites, or plant diseases. It is always advisable to contact suppliers of the products to be used as they would have the most recent side effects data.

Pests and their associated natural enemies occur on plants whether grown as arable crops (Figure 2.1), as semipermanent crops (soft fruit and orchards) (Figure 2.2), or protected within a structure (Figure 2.3). These growing systems vary in size, from extremely large areas of monocultured arable crops to smaller

areas such as glasshouses and private gardens. The larger scale arable crops are highly mechanized (Figure 2.4) but generally have lower levels of financial input (labor and materials) per hectare when compared with fruit production and protected crops.

Fruit production is generally on a smaller but more intensive scale in terms of labor requirements for pruning, harvesting, etc. Protected crop systems are intensively managed throughout much of the cropping cycle, frequently with additional heat and carbon dioxide inputs.

The three cropping systems discussed in this chapter are relevant worldwide. Such systems have a major effect on pest occurrence and distribution, so they are very relevant in

Figure 2.1   A large monoculture field of young maize in Alsace, France.

Figure 2.3   A modern glasshouse structure with heating and supplementary lights in West Sussex, UK.

Figure 2.2   Fruit trees in full blossom attract many insects, including beneficial species, pests, and pollinators.

Figure 2.4   Rapid and efficient, a combine harvester discharges its load of wheat to a trailer.

terms of pest control regardless of the size of enterprise. For example, a mixed cropping, large private garden with a small glasshouse, vegetable plot, and soft and top (tree) fruit can have the same pest spectrum in each of these systems as the largest farm, orchard, or protected nursery. However, with a greater variety of plants in a confined locality, there inevitably comes a wider range of pests and their natural enemies. Climatic factors are also important and, although many cereal pests occur only on cereals or grasses and are unable to survive on other plants, pests on protected crops grown in cool temperate regions may be found outside on the same or similar plants in the tropics and warm temperate climates.

ARABLE

Introduction

Arable crops that are grown in fields—often extensively (some with minimum tillage) and generally with an annual harvest—include cereals (Figure 2.5), beans, peas, potatoes (Figures 2.6 and 2.7), beet, and oilseed crops. In parts of Europe the development of "set-aside" arable land and uncultivated headlands has helped in the conservation of natural populations of beneficials as well as provided cover for "game" (Figure 2.8). Such areas are not sprayed with pesticides and have a wider range of plants, which support a more diverse fauna. The development of organic farming has also added to local

Figure 2.6   A potato crop in ridged rows in Belgium.

Figure 2.5   Ears of ripe wheat.

Figure 2.7   A healthy potato plant with exposed tubers.

ARABLE

Figure 2.8   A field edge with natural plants beside large monoculture fields to encourage beetles and other beneficial animals.

Figure 2.10   Corn poppies flowering in an unsprayed field margin to encourage diversity of wildlife. Note closer proximity of hedges and woodland.

Figure 2.9   Fields of ripe wheat in Devon, UK, with trees and hedges in the distance.

biodiversity, encouraging the contribution of natural enemies. New cultural systems go toward the provision of an optimum environment for beneficial insects where prey species are sufficient to enable their reproduction, as well as minimal hazards such as the use of harmful pesticides. They may also provide suitable overwintering sites within striking distance of the host plants (Figure 2.9).

From both an ecological and economic point of view, arable crops differ from orchards and protected crops in several important ways as far as their value as a haven for beneficial insects and spiders is concerned. Arable crops have considerably lower gross margins than the others, so the costs of inputs per hectare are relatively low. This is the most important influence on the management of beneficial insects and it is why the approach to IPM is so

different from that in high-value horticultural crops.

Arable Production and Biologicals

Arable crops are grown extensively over very large areas, often without windbreaks or hedgerows. Favorable conditions for the overwintering of beneficial organisms may not therefore occur. In the spring, the distance from overwintering sites is important as many beneficial species, particularly carabid beetles, overwinter in the soil beneath hedgerows and woodlands (Figure 2.10). A critical requirement for overwintering sites is the provision of the sugars, nectar, or insect honeydew required to enable flight activity, particularly in spring. If beneficial insects are slow to distribute within a crop, pest numbers can increase in their absence and growers are more likely to apply a protective pesticide as a routine measure. This is particularly so with winged aphids in cereals, as they can arrive in large numbers and may risk the introduction of barley yellow dwarf virus. There is then considerable pressure to spray a pesticide to control them and, in doing this, destroy the beneficials.

The Challenge of Biocontrol

Beneficial insects are expensive and establishing them in a crop relies on close crop monitoring to ensure that sufficient prey are

Figure 2.11   A tractor-mounted sprayer treating a flowering field of canola or oilseed rape on a spring evening.

present to guarantee that a single introduction will succeed.

Pesticides that are compatible with beneficial insects tend to be very specific in the target pest or disease that they control and are often more expensive than broad-spectrum pesticides (Figure 2.11). The introduction of commercially raised beneficial insects and the regular use of compatible and probably expensive pesticides, together with the necessary management time required to monitor the predator–prey relationships, is generally too costly to be economically viable in most arable crops.

Many countries collect data from farmers and growers to produce tables relating to crop production. In the UK, the Department of Environment, Food and Rural Affairs (DEFRA) publishes an annual survey covering the previous 11 years' data. These reports are freely available under the heading "Pesticide Usage Survey Report" and cover all agricultural and horticultural cropping including grassland for animal grazing. This consistently indicates that fungicides are the most extensively used pesticides on most British arable crops. None of the currently registered fungicides have direct insecticidal properties, although some may show negative side effects to beneficials. Where an insecticide is applied, it is usually a synthetic pyrethroid for the control of aphids. Pyrethroids are the cheapest group of agrochemicals and are therefore generally the most widely used on arable crops. They have a short harvest interval, so crops can be harvested almost immediately after application without harm to the consumer.

Synthetic pyrethroids at their commercial application rate are harmful to most beneficial insects and are likely to kill more than 75% of any with which they come into contact at the time of spraying. More importantly, this harmful effect is likely to last on the sprayed leaf for up to 12 weeks after the day of application. Unfortunately, pyrethroids do not have such a persistent harmful effect on the pests and may need to be sprayed more than once. This will prevent the successful migration of beneficial insects into the sprayed area and delay their impact as biological control agents for the target pest. The registration requirements for many broad-spectrum insecticides include unsprayed margins, or buffers, as mitigation to protect the adjacent habitats, including hedgerows, from contamination by spray drift.

The Use of Pesticides and Beneficial Insects

It has been estimated that most of the 25,000 species of terrestrial arthropods in the UK exist in farmland (Aldridge and Carter 1992), and over 99% of these are natural enemies of crop pests, pollinators, or food for other farmland fauna (Brown 1989).

In an unmanaged natural ecosystem, the natural enemy populations usually build up after the peak in the prey (pest) population. The delay in buildup can allow some pest damage to occur early in the life of the crop. The delicate balancing act that the grower faces is to decide whether this damage is economically significant enough to warrant spraying or whether it is safe to delay spraying and allow the beneficial insects to build up and to control the pest. However, if the natural enemies are not present in sufficient numbers to prevent economic damage, the grower may consider their presence irrelevant in the crop protection scheme and therefore spray.

For many years regulatory authorities have monitored and funded various research projects to determine the environmental impact of pesticide usage in arable crops. The Boxworth Project in the 1980s (Greig-Smith 1991) examined the effects of intensive pesticide use in winter wheat on populations of small mammals, birds, invertebrates, and plants. After monitoring over a 5-year period, this work indicated that insects and spiders were particularly vulnerable to pesticide usage, while some species of carabids (ground beetles) and springtails disappeared altogether.

Cereal crops sown early in the autumn may be sprayed with a pyrethroid for the control of aphids. The nontarget fauna that could be exposed to pesticides at this time are the autumn- and winter-active carabids (*Bembidion obtusum, Nebria brevicollis, Notiophilius biguttatus*, and *Trechus quadristriatus*) and staphylinids, springtails, mites, and lycosid and linyphiid spiders (Cilgi and Vickerman 1994). Ironically, carabids, staphylinids, and spiders are important predators of aphids, the target pest of the insecticide sprays. Carabids are also very susceptible to slug control chemicals such as methiocarb. Again, carabids are natural enemies of slugs and, while slug pellets protect plants from slug damage, they also kill the natural enemies of slugs.

Studies have shown, however, that spiders, staphylinids, and carabids are able to recover from autumn- and winter-applied pesticides by the following summer (Vickerman 1992). However, further applications of pyrethroids in the summer would prevent this resurgence.

Monitoring Levels of Beneficial Insects

In order to assess the potential contribution of natural enemies, they must first be captured and identified. Choosing the appropriate sampling method, time of year, and time of day when the sampling is conducted relies on knowledge of the life cycle and behavior of the natural enemy.

CARABID BEETLES

The ecology of carabids has been studied extensively, and there is sufficient information about the life cycles of the most abundant species to determine when and how to capture them. For example, some carabids, such as *Platynus dorsalis* (= *Agonum dorsale*) and *Bembidion lampros,* are characteristically ground active, which means that they can be caught in pitfall traps placed in the soil (Figure 2.12). These consist of a container without a lid or lip that is sunk into the ground so that the mouth is level with the soil. The open mouth should be protected to exclude rain and other large animals, such as toads, from falling in without preventing insects from entering. Preferably, the ground should fall away from the mouth of the container to reduce the risk of flooding. Traps should be examined daily or contain a liquid preservative (e.g., 50% ethylene glycol), since many carabids are cannibalistic (Figure 2.13).

Window traps are used to monitor the movement of flying carabids. These are vertical pieces of clear, rigid Perspex supported at various heights above the ground that obstruct

Figure 2.12   A pitfall trap to sample populations of small, ground-dwelling arthropods in arable crops.

Figure 2.13   A Petri dish with the catch from a pitfall trap.

Figure 2.15   D-vac mechanical vacuum sampler used to collect small arthropods—in this instance, in a ripe wheat field.

Figure 2.14   A plant saucer painted yellow and filled with water and a few drops of detergent is used to sample hoverflies in wheat field.

the flight of beetles. They collide with it and fall into a trough of preserving fluid below the window. The traps should be placed strategically in natural flight paths, such as between trees. Alternatively, sweep nets that consist of strong metal hoops about 40 cm in diameter, with a 30 cm handle and a calico collection bag fixed to the hoop, can be used to collect beetles in the canopy. The net is swept to and fro in front of the collector while walking through the crop or area to be monitored. Static traps, consisting of yellow painted saucers filled with water and a few drops of detergent, are used to catch several flying insects, such as hoverflies (Figure 2.14).

The beetle *Demetrias atricapillus* climbs to the upper leaves and into cereal ears and is more likely to be caught in a sweep net or suction sampler (Figure 2.15) than a pitfall trap. Suction traps usually consist of a motorized vacuum tube that can be placed over plants or the ground to lift surface-dwelling insects, mites, and spiders and deposit them in a collection container. The larvae and adults of *D. atricapillus* are mostly nocturnal. Hence, the activity of the natural enemy should be known in order that samples be taken at the appropriate time of day for a nocturnal or diurnal beetle. Suction trap sampling and sweep nets will also collect other flying natural enemies of aphids such as lacewings, hoverflies, and parasitoid wasps.

STAPHYLINID BEETLES

These can be predators, fungivores, or detritivores and can occur in very high numbers in arable crops. The ecology of many species is not known. The smaller staphylinids can be quite difficult to identify and they have often been overlooked or lumped together at the subfamily level in agricultural field studies. Like carabid beetles, staphylinids are easy to sample in arable crops using pitfall traps (see Figure 2.12). While pitfall traps are a useful and cost-effective way of determining the beneficial arthropod fauna of particular fields, it is worth noting that they only sample mobile species and that catch size is strongly affected by the weather conditions. Certain staphylinind species (e.g., *Tachinus signatus* and *Stenus clavicornis*) appear to be particularly susceptible to pyrethroid insecticides.

LINYPHIID SPIDERS

These are perhaps one of the most important predatory groups in arable ecosystems (Nyffeler and Benz 1987). Despite their potential to support a rich spider fauna, individual fields have been found to show considerable variation in the numbers of individuals and species present. Toft (1989) showed in the UK that, in some arable fields, spiders were as abundant as in natural habitats, whereas in the US Nyffeler, Sterling, and Dean (1994) reported sparse spider populations from agricultural fields. Samu et al. (2001) found that spider populations in Hungarian cereal fields were dominated by five species (*Pardosa agrestis, Meineta rurestris, Oedothorax apicatus, Pachygnatha degeeri*, and *Tibellus oblongus*). These dominant species were found in every field sampled in different regions of Hungary.

While the exact species composition may differ between countries, a similar pattern has been observed throughout Europe (Luczak 1975). The most common linyphiid spiders found in arable crops in northern Europe include *Erigone atra, E. dentipalpis, Lepthyphantes tenuis, Bathyphantes gracilis, Meioneta rurestris,* and *Oedothorax fuscus* (Sunderland, Fraser, and Dixon 1986; Topping and Sunderland 1992). Their sampling and identification are relatively easy and they can be extracted from crops by the use of a portable "suction sampler," which sucks them from the canopy (see Figure 2.15) into a collection bag, as well as by using pitfall traps. Linyphiid spiders are often caught in large numbers in arable crops in the autumn and winter (Sunderland et al. 1986). Some of these spiders can also be reared in the laboratory, which opens up the possibility of augmenting field populations if "low-tech" farm rearing methods are developed. Spiders have a high dispersal ability using wind currents, which makes them a valuable natural enemy in the large fields that are characteristic of arable systems.

Integrated Crop Management (ICM)

There are now strong commercial pressures in many countries to reduce inputs in arable crops. This has led to the development and evaluation of less intensive and more ecologically sustainable crop husbandry methods. The success of such systems depends upon their being economically feasible. While crop protection is just one aspect of crop production, both ICM and organic systems require a better understanding and the effective use of beneficial organisms in order to reduce reliance on pesticides. Recognition of both the pests and the beneficials is the first step to a more sustainable farming future.

The future of farming lies in maintaining profitability. A factor in profitability might be the use of environmentally sound systems, such as ICM or organic production. Such systems would integrate the use of genetic resistance and chemical, biological, cultural, and physical controls to reduce crop protection inputs. In the future, the genetic manipulation of crops and pests and biologicals might also be a factor.

Recognizing the potential value in the marketplace of such systems, research into integrated production methods in Europe has made rapid progress in recent years. In some countries farmers and growers are able to have their environmental farm policy and pest control strategies accredited by independent audit. This may be part of the marketing requirement to enable producers to supply directly to the large multiple retail outlets (supermarkets). In the UK, the audit is done through the auspices of LEAF (Linking Environment and Farming).

This organization provides technical assistance, through close association with the Farming and Wildlife Advisory Group, to develop wildlife conservation plans on farms and accredit the environmentally sensitive farming methods employed. In support of these developments, the Countryside Stewardship Scheme offers payments to farmers; these further encourage conservation plans as part of the land management program on a farm.

Conservation Strategies Using Headlands and Beetle Banks

These strategies are a valuable tool in ICM strategies and are a less expensive means of introducing natural enemies into a crop, especially where low gross margins inhibit the mass release of commercially raised beneficial insects. Research on "conservation headlands" has shown the benefits of a 6 m wide strip between the crop edge and the first tramline on the survival of beneficial organisms. Beetle banks, which are grass ridges, can be positioned in the middle of large fields to form refuges for spiders and staphylinid and carabid beetles. This enables predators to overwinter more effectively in midfield refuges (Figure 2.16), from where they can spread into the crop in the spring.

Figure 2.16   A grassy "conservation headland" or beetle bank between large open fields, unsprayed strips form refuges for beetles and spiders.

FRUIT

Introduction

The natural parasites and predators found on fruit crops are surprisingly similar, irrespective of whether the crop is strawberries (Figure 2.17) or cranberries in the United States or apples (Figure 2.18), citrus (Figure 2.19), or grapes (Figure 2.20) in Europe. The same predators and parasitoids feature as antagonists of aphids and mites, scale insects, and leaf rollers (tortrix moth larvae) in almost all crop systems and in various countries.

Figure 2.17   Ripe, field-grown strawberries in Berkshire, UK.

Figure 2.18   Ripe Cox's EMLA clone apples on the tree in Oxfordshire, UK.

Figure 2.19   Flowers and ripe fruit of Navelina oranges on a tree near Valencia, Spain.

Figure 2.20   Mature pinot noir grapes on the vine in an open-field vineyard, Champagne, France.

The same pest species also feature in different crops and on different continents. Although there are often geographical differences as to which are the relevant species, it is usually insects in the same genus that attack key pest species. For this reason many of the species descriptions given in this book are equally applicable throughout most temperate regions of the world.

Figure 2.21   Flowering cordon apple trees in the Loire Valley, France.

Fruit Production and Biologicals

Fruit crops can range in size from trees in a small plot to a hundred or so plants or up to many hectares of continuous culture. Plant size also varies from well spaced, large, mature trees to densely planted, multirow blocks of small plants. The ground under the plants may be bare earth, neatly mown grass, or mixed vegetation. The margins and boundaries may be old, diverse hedgerows; coniferous or deciduous windbreak trees; mechanical windbreaks; or simply a barbed wire fence. With such a range of possibilities, it is difficult to generalize about fruit crops. While many pest species will be attracted to orchards irrespective of their configuration, their natural enemies may be more discerning. Immaculate rows of small trees with bare earth beneath them (Figure 2.21) will present few hiding places and little alternative food for voracious predators. That does not mean that weed-free crops are inherently unsuitable for beneficial arthropods. However, if pest numbers fall within the crop, then the predators may leave altogether rather than move to nearby food on other plants.

It is not usually practical to buy and release beneficial arthropods in fruit orchards.

Extremely large numbers would be needed and they would be unlikely to be seen again. The key to establishing biological control in commercial fruit crops is to encourage and enhance populations of the naturally occurring predators and parasitoids.

Predators in fruit crops fall into two distinct types: residents and colonists. The resident species are present throughout the whole of a growing season and are established and well placed to feed on the early pest individuals as they arrive in an orchard. Residents tend to be nonflying and polyphagous and will occur whether or not there is an apparent source of prey. Two important residents in fruit orchards are often earwigs and spiders. Although earwigs may be considered to be a pest since they can damage the surface of fruit—particularly in crops such as peaches and apricots—they are also predators of aphids and mites and can occur in very high numbers in orchards. Since they are active at night and hidden in crevices during the day, earwigs are often overlooked.

The colonists include predatory bugs of the families Anthocoridae, Miridae, and Nabidae together with coccinellid beetles, hoverflies, and lacewings. Colonists tend to be adult flying insects that overwinter as adults and seek shelter among bark and holes in trees or in hollowed dead stems of perennial plants, from where they migrate to and from the orchard. Initially feeding on early spring flowers, they are attracted to an orchard because of the presence of prey (usually in the form of a pest outbreak). Since colonists are usually highly mobile, they can occur in extremely high numbers. For some colonists, such as hoverflies and some lacewings, only the larval life stage is predatory; for others, such as predatory bugs and ladybirds like *Coccinella septempunctata*, both the larvae and the adults are predatory.

Fruit crops can be managed so as to nurture field resident populations of natural enemies and to provide an environment where immigrant predators and parasitoids can become established and thrive. By and large, fruit growers prefer bare soil around their trees to avoid alternative hosts for specific insect and mite pests.

The Challenge of Biocontrol

Not surprisingly, the first step to enhancing naturally occurring biological control agents is to carefully examine the use of pesticides (Figure 2.22), particularly insecticides and acaricides. In the winter months applications of broad-spectrum products such as pyrethroids and organophosphates may not be damaging to beneficial populations because the majority will be hidden away in crevices or present as eggs and relatively protected in the bark of trees. Broad-spectrum pesticides can have a place in an IPM program if they are used at the right time of year. For a grower to determine the likely risk to nontarget arthropods from a product, he should consult the label and seek confirmation from the literature, websites, or an advisor as to the likely impact.

Members of the IOBC/wprs (International Organization for Biological Control/West Palaearctic Regional Section [http://www.iobc-wprs.org/expert_groups/01_wg_beneficial_organisms.html]) working group on pesticides and beneficial organisms screen new and existing products for harmlessness using "worst case" laboratory tests. Their findings are published in terms of a classification from 1 to 4 (1 being harmless with

Figure 2.22   Tractor-driven, air-assisted sprayer treating an apple orchard, Provence, France.

<25% mortality and 4 being harmful). The most up-to-date information can be found by entering the phrase "side effects of pesticides on beneficials," which leads to interactive tables produced by some of the commercial biocontrol producers. While these results are useful, they must be treated with some caution. Products found to be harmless can be considered low risk in the field, whereas those found to be harmful in laboratory tests require further evaluation to determine the magnitude and duration of any side effects that would occur in a crop system.

Although the majority of fungicides are considered to be relatively harmless to insects and mites, most organophosphate products, whatever their intended use, are insecticidal and can be very harmful to parasitoids, predators, and pollinators. Sulfur, commonly applied as a fungicide in France, is particularly damaging to populations of predatory mites such as *Typhlodromus pyri* and *Amblyseius* (syn. *Neoseiulus*) *californicus*. Since these are relatively immobile mites that can be important predators of spider mites, their presence is desirable. Very often the first signs a grower will notice when he has killed predatory mites with a broad-spectrum product will be an outbreak of spider mites that were previously being controlled. Fungicides containing the active ingredient mancozeb are not harmful to predatory mites after the first application, but typically affect them adversely after a fourth or fifth treatment in a single season. Among the acaricides a number of products are relatively selective (e.g., the growth regulators) and tend to be more effective against spider mites than against predatory mites; some have only a limited effect against the beneficial ones. Insect growth regulators do not give immediate control (often working only when the pest molts) and for this reason have sometimes been avoided by growers looking for a quick kill with a dramatic knockdown.

Where aphids are a problem, the use of selective aphicides can result in minimal disturbance of nontarget insects. When broad-spectrum insecticides are used to control a particular pest problem, it is usually preferable to select a product that is relatively short lived. Products such as chlorpyrifos are highly efficacious insecticides and will certainly kill most pest and beneficial mites, spiders, and insects in a treated crop. However, they will undergo relatively rapid chemical degradation and the leaves and fruit will no longer be toxic to immigrant predators and parasitoids after about 10 days. Larger orchards will be less readily recolonized than smaller ones.

Intelligent selection of pesticides is a critical component to optimize the role of natural enemies in pest control. While it is easy to recommend using a small number of relatively selective pesticides, it is very dangerous for IPM to become totally dependent on them. For several years, in a range of fruit crops in southern Europe, IPM relied heavily on treatments with the insect growth regulator diflubenzuron and phosalone, together with the actions of beneficial arthropods. But eventually the pests became resistant to these two products. Faced with uncontrollable pests, the growers resorted to broad-spectrum products. While they were aware of the value and importance of natural enemies, getting a satisfactory crop is of paramount importance.

Unfortunately for growers, the pests may also develop cross resistance to many products, particularly in populations of the codling moth, *Cydia pomonella*. In parts of southern France the codling moth is now unaffected by many insecticides and resistance is a serious problem. Repeated applications of broad-spectrum organophosphate and synthetic pyrethroid products have devastated populations of virtually all the main biological control agents. Interestingly, some predators in a few localities have also been found to be resistant to organophosphates. The key point to arise from this is that IPM strategies should not rely heavily on only one or two types of modes of action pesticides for the control of pests. Fortunately the commercialization of *Cydia pomonella* granulosis virus (*Cp*GV) products offers a more IPM-friendly approach, although tolerance to some isolates is a threat that must be considered.

The use of smart methods of control that affect only the pest species, such as mating-disruption techniques and baits with pheromones (attract and lure), is preferable to spraying and these methods are becoming more widely used. Careful choice of crop protection products is critical to maintaining biodiversity within fruit crops. IOBC/wprs guidelines for integrated production of pome fruits in Europe (Cross 2002) state nine criteria that should be taken into account when considering the suitability of a product for use. These include toxicity to man, toxicity to key natural enemies, toxicity to other natural organisms, pollution of ground and surface water, ability to encourage pests, selectivity, persistence, incomplete information on any of the above, and the necessity for use. Based on these criteria the IOBC/wprs subgroup for integrated fruit production considers pyrethroid insecticides and acaricides, non-naturally occurring plant growth regulators, organochlorine insecticides and acaricides, and toxic water-polluting or very persistent herbicides to be incompatible with integrated fruit production.

The fruit crop habitat itself can be enhanced to encourage predators and parasitoids. Early in the season, before any pest species have been observed on the fruit trees, quite large numbers of predators, particularly Heteroptera such as *Orius* or *Deraeocoris* spp., have been observed on weed species in the margins or in the understory of orchards that are not being intensively managed. As the season progresses these species are found on the trees themselves,

feeding on spider mites, aphids, and psyllids. Such predators often overwinter beneath bark or in dead leaves and appear to favor a sheltered, protected environment from which to extend their activity up into the trees. The provision of areas of "weeds" in the margins of orchards or allowing a mixture of plant species to grow underneath the fruit trees will definitely encourage predators.

However, it should be noted that many pests have alternate host plants and that the wrong weeds could encourage pests. For example, the common weed *Plantago lanceolata* (Figure 2.23) (ribwort, buckhorn plantain, lamb's tongue, etc.) is an alternate summer host for rosy apple aphid *Dysaphis plantaginea* (Figure 2.24). If pesticides are applied to an orchard, then the understory vegetation or refugia will provide shelter and perhaps

Figure 2.23 *Plantago lanceolata* **(numerous common names, including ribwort, buckhorn, English plantain, or lamb's tongue) is an alternative summer host for rosy apple aphid.**

Figure 2.24   Colony of rosy apple aphid (*Dysaphis plantaginea*).

Figure 2.25   A honeybee (*Apis mellifera*) on apple blossom.

unsprayed reservoirs from which predators can recolonize trees. The addition of flowering plants both within orchards and at their margins has been shown to increase predator abundance, particularly of spiders. However, when these plants are flowering, care must be taken to avoid spraying any pesticides that would be toxic to honeybees (Figure 2.25) or bumblebees.

"Functional biodiversity" is the current term for allowing a wider range of plants and associated insects to flourish within fruit crops. It is impractical for growers to identify all the plants and invertebrates within the crop, so attention is now focusing on identification of one or more bioindicators (Brown 2001) or biomarker guilds. These guilds are made up of an association of species known to react

to impacts or changes within the ecosystem (Paoletti and Bressan 1996). Concern over "bioindicator" terminology and its meaning was discussed by van Gestel and van Brummelen (1996). Bioindicators will allow managers to determine the health of a system without needing to survey and identify all the invertebrates.

An additional step is to provide a source of prey early in the growing season—for example, by introducing plant species that are attacked by their own aphids or mites, particularly early in the year. As long as the plants are carefully selected, their aphids will be specific and will not feed on the fruit trees. However, many of the aphid predators, such as hoverflies, coccinellids, and lacewings, may be attracted to the orchard by the aphids on the so-called weeds, but they will feed equally well on the aphids on the trees. If flowering plants are also present in the orchard, then these will provide a source of pollen (which can serve as a secondary food for predatory mites and predatory bugs such as *Anthocoris* spp.) and will attract hoverflies, the larvae of which are predators of aphids.

As well as the area of the crop itself, the margins and boundaries can be managed to make them attractive to beneficial species (e.g., by planting appropriate tree species for oviposition by predatory bugs). The neighboring crops and habitats will also strongly affect the overall abundance of predatory and parasitoid species in a given location. Boller (2001) gives examples from Swiss vineyards, where protection of hedges and borders in close proximity to the crop is encouraged and where blackberries (*Rubus fruticosus*) provide a source of predatory Phytoseiid mites (Baur et al. 1998), and wild roses (*Rosa canina*) provide a source of host insects for the leafhopper parasitoid *Anagrus atomus*. Stinging nettles (*Urtica dioica*) are tolerated since they are a food plant for many nonhost plant-specific insects such as Lepidoptera and are important for building up their parasitoid complex.

Orchards backing onto woodland, grassland, or simply other orchards that are rich in insects, spiders and mites will be likely to have colonists waiting to move in and begin feeding on pests. Unfortunately, past subsidies paid in Europe to growers to remove apple orchards resulted in destruction of many of the old and relatively unintensive orchards, which harbored high populations of virtually all of the beneficial groups. In part of the Ardeche region of France, apples and pears are grown in a patchwork of small orchards, often with vines in adjacent blocks. The fruit is managed so as to encourage predators, which can move freely from one orchard to another. Care is taken when treating the vines so as not to overspray the orchards.

The sex pheromones of many lepidopteran pests of fruit crops, including soft fruits, have been identified. Semiochemicals, attractants, and pheromones can be used in sampling to determine timing of pesticide application, used in lure and kill baits as an alternative to spraying, and used to disrupt male mating behavior. Cross and Hall (2009) describe developments in discovering the sex phero-mones for three important pests of soft fruit: the European tarnished bug (*Lygus rugulipennis*), strawberry blossom weevil (*Anthnomus rubi*), and the apple leaf midge (*Dasineura mali*). The mating disruption technique is based on the release of a synthetic version of the natural pheromone, either from a dispenser or by widespread broadcasting. Dispensers are placed in the orchard in the spring before the flight of the males of the target species. This causes a reduction in the target insect pop-ulation by masking natural olfactory traces and reducing the number of viable matings. Successful area-wide mating disruption has occurred in Europe with the codling moth, *Cydia pomonella*, and the grape berry moths, *Eupoecilia ambiguella* and *Lobesia botrana*. Aerial dissemination of synthetic pheromones was used to control tortricids in 50,000 ha of orchards and vineyards in 2001.

Pheromones are highly volatile and often rapidly degraded by UV light, necessitating frequent reapplication. Mating disruption is vulnerable to wind and is affected by the relief of the land. Mating disruption works best in large, flat orchards with smaller trees (<4 m tall) of the same height. In this situation it is easier to use a number of dispensers to generate a relatively homogeneous cloud of pheromone.

Major Pests
In the 1980s spider mites and aphids were probably the major pests in fruit orchards in Europe, with psyllids, scales, codling moth, and leaf rollers occurring as minor pests. In low-input orchards in southern Europe it was not uncommon to see major outbreaks of fruit tree red spider mite, which in turn attracted very high densities of predators such as the ladybird beetle (*Stethorus punctillum*). Aphids, such as the woolly apple aphid (*Eriosoma lanigerum*), would appear in large, white, waxy colonies and attract high numbers of the species-specific parasitoid *Aphelinus mali*. In recent years the codling moth and leaf rollers have increased to become the major pest species.

Integrated Crop Management
Growers often believe that the actions of one or two key species are responsible for all bio-logical pest control in a commercial orchard. Classical examples of biological control tend to reinforce this hypothesis with well-known cases of one antagonist against one pest. In commercial orchards, different predatory or parasitoid species tend to become the main antagonists in different growing seasons. Sometimes this is attributable to weather con-ditions or to the availability of suitable prey, which will favor one particular species over another.

Looking at populations of pests, predators, and parasitoids in fruit crops, it seems that it is the combined efforts of all the beneficial arthropods that may actually limit or contain pest outbreaks. Mites, aphids, and psyllids can be attacked by a diverse range of different predators and parasitoids, some of them generalists and some of them specific antagonists of individual pest species. At any time a single tree will contain a mixture of beneficial arthropod species depending on the time of year and the abundance of prey. For example, trees in a low input pear orchard in Provence, France, sampled in May 1996 (Figure 2.26) contained, on average, 250 predators and parasitoids per tree. Of these there were about 100 predatory bugs, 65 earwigs, 25 spiders, 17 ladybird beetles, 16 parasitic wasps, 10 lacewings, eight hoverflies, and nine representatives of other predatory groups. This orchard had a small outbreak of pear *Psylla,* which was clearly the primary food for the dominant bug species. The species composition and the relative abundance of different beneficial groups will differ between orchards in the same region in the same season and will also vary considerably between years.

In all field crops and in fruit crops in particular, there is no one superpredator. The species that are bred and sold commercially tend to become well known and these usually represent the most important predatory and parasitoid groups. However, it is sometimes the lesser known insects, such as the odd looking bug, *Heterotoma planicornis,* or perhaps action of unfashionable but purely

predatory groups such as spiders, that end up limiting pest populations in a particular orchard.

Aphids and mites are external feeders on leaves and shoots of the trees and are readily accessible to predators and parasitoids. In contrast, the larvae of codling moth quickly penetrate the apple or pear fruit and are protected from predation by the fruit itself. Growers cannot sell fruit with the blemish of an entry hole and would strive for very high levels of control. For this reason and because this moth (Figure 2.27) can cause major damage even when present at low densities, chemical control has been the main method employed against codling moth. Since the period of egg laying and egg hatch may last for 3 or more weeks, it is often necessary to make three to four applications against the spring generation. Classical control of codling moth was attempted in New Zealand by introduction of parasitoids such as *Ascogaster quadridentata* and *Liotryphon caudatus* from overseas. Although these parasitoids established successfully on the high populations of codling moth in neglected apple orchards, they only made a minor contribution to control in commercial orchards. Similarly, biological control of codling moth in California using the pupal parasite *Mastrus ridibundus* was not able to achieve an acceptable level of control in apple orchards.

Figure 2.26   Sheets beneath fruit trees used to collect and monitor an insect population.

Figure 2.27   A codling moth (*Cydia pomonella*) caterpillar inside a gallery in an apple.

PROTECTED CROPS

Introduction

Protected environments of the glass or polythene house (Figure 2.28), conservatory (Figure 2.29), or similar structures generally provide a very favorable environment for plant growth (Figure 2.30). In doing so, plants grow faster and pests or diseases can have a favored environment in which to thrive. Pests that naturally occur only during summer months may remain active throughout most of the year (e.g., various moths, such as the carnation tortrix, will breed continuously in a heated structure), whereas in unheated structures pests may not be active during the winter months but they will be seen earlier in the growing season. At this time the protective structure not only reduces the effects of adverse weather conditions but also prevents many migrating pests from entering. This can work in growers' favor, particularly when introducing beneficial organisms early, as the organisms are less likely to escape and more likely to control pest infestations.

Crop Production and Biologicals

Once equilibrium has been achieved between pest and natural enemies, it becomes easier to control subsequent pest outbreaks and the whole control program becomes more stable. The idea of "seeding" plants with a pest and later beneficial insects to create a bank of biological control agents was widely adopted in the early 1970s and is known as the "classical" introduction method or "banker plant system" (Hussey, Read, and Hesling 1969). A modern variation on this theme is to use pests that are specific to a group of plants but that can sustain natural enemies capable of living off other pests. Trays of barley, wheat, or pots of maize infested with cereal aphids are exposed to parasitoids or predators (*Aphidius* or *Aphidoletes* spp.). When the host pest numbers are controlled and beneficials are in abundance, the plants are brought into the glasshouse, where the natural enemies disperse to find other aphids to attack. Cereal aphids, being

Figure 2.28   Commercial polythene houses near Almeria, Andalusia, Spain.

Figure 2.29   Conservatory with diverse collection of plants; some may remain for several months or years, allowing a range of pests to establish on plants and within the structure.

Figure 2.30   A commercial poinsettia crop "Spotlight" (*Euphorbia pulcherrima*) in West Sussex, UK.

PROTECTED CROPS

host plant specific, are unable to survive on protected crops, so they pose no threat. This system works well in early planted cucumber crops where the principal aphid is *Aphis gossypii,* which has an extremely high rate of reproduction. The banker plant system continues to produce useful numbers of natural enemies for several weeks, usually long enough for good establishment of parasitoids or predators through the crop.

An alternative method is to lightly "seed" known or tagged plants with a few pest organisms, leave them for a week or so, and then introduce their biological control agent directly to the preinfested plants. This latter method is gaining popularity on protected sweet pepper crops (Figure 2.31), which can be seriously damaged by spider mites during the heat of summer when most predators struggle to maintain control.

Pest Monitoring

Monocrops are very vulnerable to epidemic attack—largely because of the close proximity of large numbers of identical plants (Figure 2.32). When these and environmental conditions are suitable for the pest, epidemic development occurs. In contrast, mixed crop plantings generally have fewer overall problems but can suffer from "hot spots" mainly due to some plants being more susceptible (or flavorsome) to certain pests. For example, by experience most growers are aware of their most susceptible plant cultivar. Prompt action in both situations can prevent a lot of damage.

Figure 2.32   A commercial strawberry crop, "Elsanta," suspended above the ground for improved access, light transmission, and ease of picking in West Sussex, UK.

Figure 2.31   Ripe orange colored sweet peppers (*Capsicum annuum*) in a commercial organic glasshouse, Somerset, UK.

Figure 2.33   Sticky yellow traps hang above a tomato crop to monitor and help control adult whitefly population; colored strings on bobbins indicate double heading of plants (one root system with two growing heads).

Crop monitoring by visual assessment of leaves and plants is the most accurate method of detecting a pest organism. However, colored sticky traps (Figure 2.33) are routinely used to monitor flying pests (Sunderland et al. 1992). Yellow traps are used for alate (winged, flying) aphids, adult leaf miner flies, adult thrips, and whitefly; blue are more selective, generally attracting adult sciarid fly and thrips, and orange are used for adult carrot fly. The addition of a sex attractant pheromone makes them even more effective. Sticky traps placed at the top of a crop canopy will reduce the pest population quite rapidly. In a small greenhouse they may suffice for the whole season's pest control.

Several natural enemies are available and most are commercially mass reared from individuals collected within the country of use, so they do not require registration for release. Many beneficial insects are more or less host specific and attack several species within the same genus. Several aphid parasitoids have a limited host range, usually within similarly sized aphids, independent of the host plant. Predators tend to be more polyphagous and attack a much wider host range. *Macrolophus pygmaeus* (Mediterranean origin and licensed for release in several countries) will live successfully on whitefly, leafhopper, leaf miner, spider mite, moth eggs, and young caterpillar and, in the absence of suitable prey, can survive on plant sap. Alone they can reduce a pest population to below economic damage levels or in some instances eliminate a pest completely. This particular predator can also cause considerable yield losses on cherry tomato plants by feeding on pollen within the flowers after its prey has been consumed. It is then, at certain times of the year, regarded as a major pest.

The Challenges of Biological Control
Glasshouse biological control started when the whitefly parasitoid *Encarsia formosa* was first used commercially in the late 1920s. It was found in a tomato crop and its use was exploited following research into breeding methods done at the Cheshunt Experimental Station in Hertfordshire, UK. During the 1930s it was mass produced and exported to several countries around the world. Further development of biological control all but ceased following the discovery of *Encarsia*. This was due to the Second World War, the subsequent widespread use of organochlorine pesticides such as dichlorodiphenyltrichloroethane (DDT), and discovery of organophosphate insecticides. These were broad-spectrum, long-persistence products that initially controlled most pests for several weeks or months after a single application.

Pest resistance and environmental concerns led to a resurgence of interest in biological control methods. The spider mite predator *Phytoseiulus persimilis* was first used in 1960 when, in Germany, Dr. Dosse found it with spider mites on a consignment of orchids from Chile (Bravenboer and Dosse 1962; Hussey 1985). *E. formosa* and *P. persimilis* were once again used in research programs and became available commercially to tomato and cucumber growers. The invention of synthetic pyrethroids in the early 1970s and their widespread use by the end of that decade again slowed development of natural enemies. However, reports of whitefly becoming resistant to massive doses of pyrethroid insecticides soon brought biological control back in demand.

Minor pests, which had previously been easily controlled by pesticides, now became more of a problem. The lack of selective pesticides (those that kill a pest without undue harm to a beneficial organism) led to a wider range of natural enemies being used. The mass production of new beneficial insects required new techniques and methods to distribute them and integrate their use. Methods of crop culture changed and soil-borne pathogens were largely controlled by growing in soil-less systems. The use of peat, mineral wool or stone wool, and hydroponic systems resulted in a change in the importance of some pests.

PROTECTED CROPS

Cucumbers were traditionally grown on soil ridges and then straw bales—environments generally too wet for thrips to build up to seriously damaging levels as they pupate on the ground. When the crop was produced on a much drier polythene-covered floor, (Figure 2.34) thrips became a major pest problem.

The predatory mite *Amblyseius cucumeris* (syn. *Neoseiulus cucumeris*) was known about for many years and indeed frequently entered crops naturally. However, with the change in growing practices there was a need to introduce many thousands at one time to achieve adequate thrip control. Thus, combined efforts by research teams and commercial producers developed the controlled release system (CRS) (Figure 2.35) for use on cucumbers and other crops. Small sachets contain a breeding population of *A. cucumeris* and a source of food in the form of a stored food mite such as *Acarus siro* (flour mite), *Carpoglyphus lactis* (dried-fruit mite), or *Tyroglyphus* spp. (mold mites) are placed one to a plant and release predators over a 6- to 8-week period. Subsequent research has shown that these sachets can be successfully used to produce a range of predatory mites to control and prevent the establishment of several pest species on several crops, including ornamental plants, where their use on hanging baskets is extremely effective.

New control agents and methods to use them continue to be developed, while new selective pesticides take much longer and cost considerably more. Insect parasitic nematodes, which normally inhabit the film of water that surrounds soil particles, may also be applied as a spray to foliage with the water deposit, allowing the nematodes to swim over the surface (Piggott et al. 2000). In this way parasitic nematodes can be used as a living

Figure 2.34   Healthy commercial cucumber crop in Hertfordshire, UK.

Figure 2.35   An *Amblyseius* controlled release system (CRS) sachet.

pesticide to control leaf miner larvae within the leaf and scale insects and thrips on the leaf.

Integrated Crop Management

Most naturally occurring beneficial insects are only active during the summer months; they overwinter in a protected environment and reappear when temperatures and daylight hours increase. This usually corresponds to several weeks after the pest species have started their reproductive cycle and sufficient prey exists for their survival. Therefore, if biological control agents are to be used successfully, they must be provided with the appropriate conditions: an average temperature of 14°C (57°F) or above (lower temperatures will not kill but will slow their development), sufficient hours of light to maintain activity, and the pest organisms in low numbers. In heated crops, most IPM programs start almost immediately when the crop is planted; however, without supplementary heat, the start may need to be delayed until early spring.

In most situations more than one pest occurs at any one time and there are also plant diseases, weeds, and nutritional disorders that may need to be controlled. IPM is the integration of biological control agents with selective pesticides and cultural control techniques. In addition, ICM includes all aspects of crop production as well as the choice of cultivars with pest or disease resistance, the appropriate use of fertilizers, and manipulation of the environment where appropriate.

Cultural control includes polythene-covered floors to reflect light to the crop (Figure 2.36) and also prevent weeds or volunteer plants that may require herbicide treatment. It also includes the use of sticky traps to monitor

Figure 2.36   Young tomato plants growing in sterile artificial media (stone wool) on white polythene-covered floor to reflect light and protect from pests in West Sussex, UK.

Figure 2.37   Last fruit of the season from a soil-grown tomato crop in a polythene tunnel in Portugal; the leaves are removed to aid ripening and reduce pest and disease pressure.

and help control flying insects. Leaf removal is a useful technique in reducing pest populations and aids ripening and picking of fruit (Figure 2.37). IPM techniques are available to amateur and professional growers alike and can work extremely well. The majority of cucumber, sweet pepper, and tomato crops produced in protected environments and an increasing area of ornamental crops use full IPM programs to control all the pests.

Biology of some common target pests

INTRODUCTION

The organisms included here are representative examples of the major pest species found throughout the world that are the hosts of the beneficial organisms that feed in or on them. In most instances, a natural enemy is closely associated with its host pest, so this chapter should aid the identification of both. The pests are described with the damage they cause (symptoms), their life cycles, and an indication of the organisms that may be found attacking them (beneficials). Pests feed on plant or animal tissue, causing damage that, in many instances, results in economic crop loss.

In most environments there are natural enemies that use the pest species as part of their diet and usually reach a population that is in equilibrium with the pest population. Such equilibrium results in reduced pest damage while leaving enough of the pest population to support the natural enemies. However, in some circumstances, pests may be found in an alien environment, such as on houseplants growing in offices or shopping malls; these pests may not normally be present locally but are characteristic of the plants and their environment.

It is now possible to initiate and control many pests effectively by introducing mass-produced biological control agents, often in place of or integrated with chemical pesticide sprays.

There are numerous books available that can give far more detail on the pest species found damaging crops (see "Further Reading" section at the end of the book). In this chapter, just some of the more common pest organisms that may be encountered are described (Table 3.1).

COMMON PEST SPECIES

Slugs and Snails

Land mollusks are not insects; they are traditionally divided into two categories. Slugs and snails are gastropods; many are in the order Pulmonata—so called because they breathe air directly with a lung. Those without an external shell or with one that is very small in relation to the body are slugs (Figure 3.1), and those

TABLE 3.1 Taxonomic classification of the major pests illustrated in chapter 3

Phylum and subphylum (SP)	Class	Order	Family	Common name	Page
Mollusca	Gastropoda	Pulmonata		Slugs and snails	33
Arthropoda	Arachnida	Prostigmata	Tetranychidae	Spider mites	36
		Trombidiformes	Tarsonemidae	Tarsonemid mites	38
			Eriophydae	Big bud mite	38
(SP) Crustacea	Malacostraca	Isopoda	Armadillidae	Woodlice	38
			Porcellionidae	Woodlice	39
			Oniscidae	Woodlice	39
(SP) Myriapoda	Diplopoda	Julida	Julidae	Millipedes	39
(SP) Hexapoda	Insecta	Hemiptera	Cicadellidae	Leaf hopper	41
			Psyllidae	Psyllids	42
			Aleyrodidae	Whitefly	42
			Aphidoidea	Aphids	44
			Coccidae	Scale insects	46
			Pseudococcidae	Mealybugs	46
		Thysanoptera	Thripidae	Thrips	49
		Lepidoptera		Caterpillars	51
			Tortricidae	Tortrix moth	51
		Diptera	Agromyzidae	Leaf miners	56
			Sciaridae	Sciarid flies/fungus flies	57
			Ephydridae	Scatella flies/shore flies	58
		Hymenoptera	Tenthredinidae	Sawfly	53
		Coleoptera	Scarabaeidae	Chafer beetles/grubs	58
			Elateridae	Click beetle/wire worm	59
			Chrysomelidae	Flea beetles	60
				Asparagus beetle	60
				Colorado potato beetle	61
			Curculionidae	Weevils/vine weevil	62

Figure 3.1   A keeled slug (*Tandonia budapestensis*).

that can retreat into a protective shell are snails (Figure 3.2). Slugs of the family Testacellidae are large (6–12 cm long when extended) and have a small shell at the rear that covers the lung. They are carnivorous and feed on earthworms. All slugs and snails have unsegmented bodies with four tentacles on their heads; the upper pair is larger with eyespots at the tip (Figure 3.3).

All mollusks are most active in moist conditions; in periods of dry, windy weather, they seek shelter and become inactive. With a decrease in light intensity, together with a fall in temperature and rise in humidity, they move onto vegetation. They feed on a wide range of living and decaying plant material and damage

Figure 3.4   A slug (*Arion distinctus*) on a severely damaged maize leaf.

Figure 3.2   A garden snail (*Helix aspersa*).

Figure 3.5   Hosta plants are readily attacked by slugs and snails, which leave shredded foliage.

Figure 3.3   The head of a slug (*Arion ater rufus*) showing tentacles with eyes on the tips of the upper pair.

appears as irregular holes (Figure 3.4) with smooth edges caused by the rasping action of their mouthparts (Figure 3.5). Plant seedlings and young plants can be chewed and killed, while older plants may be severely damaged. A slime trail (Figure 3.6) is nearly always present and may lead to the pest's resting place. Mature slugs and snails are hermaphroditic, having both female and male reproductive organs. When mating, they slide against each other and both adults may be fertilized simultaneously. Clusters of eggs are laid in soil or decaying organic matter (Figure 3.7), which provides the necessary protection from adverse conditions (freezing and desiccation). Some species living in warmer environments may have two or more generations per year;

Figure 3.6   A gray field slug (*Deroceras reticulatum*) on a damaged oilseed rape or canola leaf with slime trail.

Figure 3.7   A slug (*Arion distinctus*) with its eggs.

others living in cooler conditions may have overlapping generations that take a couple of years to mature.

Biocontrol
Natural enemies of slugs and snails include insectivorous mammals (hedgehogs and moles), several bird species, frogs, toads and lizards, parasitic flies (Sciomyzidae), predatory insects such as carabid beetles, and entomopathogenic nematodes such as *Phasmarhabditis hermaphrodita*.

Figure 3.8   Speckling damage on cucumber leaf indicating spider mite feeding on the underside.

Figure 3.9   Severe carmine spider mite (*Tetranychus cinnabarinus*) damage to rose leaves.

Spider Mites

This section details the minute plant-damaging spider mites commonly found speckling leaves (Figure 3.8) on many crops throughout the world. High numbers of spider mites can distort young plants or completely desiccate leaves (Figure 3.9) and flowers. When the infestation is severe, webbing can form, particularly on the apical growing points of affected plants. The webbing is a mass of fine silken threads in which there is a very large population of spider mites (Figure 3.10). Mites on these fine threads can be caught in air currents and moved considerable distances. They can also be transported mechanically (including by humans, on clothing, and by plant movement) and are often deposited on other plants.

Figure 3.10   Two-spotted spider mite (*Tetranychus urticae*) damage to *Phaseolus* bean leaves with active mites on webbing.

Figure 3.11   An adult *Tetranychus urticae* female on fine webbing with egg.

Figure 3.12   An adult female fruit-tree red spider mite (*Panonychus ulmi*) with an egg on an apple leaf.

Figure 3.13   Overwintering eggs of *Panonychus ulmi* on bark of an apple tree.

The Tetranychidae are the larger mites (up to 0.75 mm in length) and are usually yellow-green to dark red in color. However, when conditions are unfavorable, some species (e.g., *T. urticae*) turn a bright orange-red color and enter a diapause state in which feeding and egg production cease until more suitable conditions return. Under glasshouse conditions where good plant growth and temperatures above 16°C (61°F) can be maintained, these mites are active through the year, although their life cycle is considerably slower at lower temperatures. Female *T. urticae* produce up to 100 spherical eggs 0.14 mm in diameter (Figure 3.11); they are initially transparent but turn white to light yellow just prior to egg hatch. The eggs of the fruit tree spider mite, *Panonychus ulmi*, are orange in color (Figure 3.12); egg clusters are deposited on the bark of fruit trees where they overwinter (Figure 3.13). A six-legged larva emerges and feeds for a short while before settling on the leaf to form a protonymph. This stage may last a few days, depending on temperature, before developing into a deutonymph. At this stage females are distinguishable from males by their larger size and rounded shape, while males have a pointed rear and are much more active.

Biocontrol

Predatory mites such as *Phytoseiulus per-similis* provide good control of glasshouse spider mite (*T. urticae*) but limited control of fruit tree spider mite (*P. ulmi*). However, the fruit tree mite is well controlled by *Typhlodromus pyri*. Various species of amblyseid mite include spider mites in their diet. The midge predator *Feltiella acarisuga* is more polyphagous and will attack most spider mites. Other predators include lacewings, mirids (*Macrolophus* spp.), anthocorids, and entomopathogenic fungi.

Eriophyid and tarsonemid mites are much smaller, about 0.25 mm in length, and are easily overlooked. The more commonly found ones include bulb scale mites, cyclamen/straw-berry mites (*Phytonemus pallidus*), broad mites, and gall mites. Damage usually occurs on young growth in the form of distorted leaves and galls (Figure 3.14) that are fre-quently discolored and have a shiny appear-ance. The mites are slow moving and creamy light brown in color. Eggs are laid on the leaf or flower surface and hatch into six-legged larvae that, after feeding for a few days, enter a resting stage. Male mites may be seen carrying a resting stage female nymph on their backs using a pair of anal claspers. The female is released when she matures and they immedi-ately mate.

The predatory mite *Hypoaspis* spp. is used for bulb scale mite control and *Amblyseius cucumeris* to control leaf- or flower-feeding tarsonemid mites. Other predatory mites are also likely to feed on these pest mites, but their efficacy has not been determined.

Woodlice

Woodlice, pillbugs (Figure 3.15), slaters, or sowbugs are all common names for this group of crustaceans. They are more closely related to crabs, crayfish, and shrimps than to insects. Some woodlice are brown with small spots (Figure 3.16); slaters tend to be a uniform slate

Figure 3.15   An adult pillbug or common pill woodlouse (*Armadillidium vulgare*); these can roll into a tight ball to escape predation.

Figure 3.14   "Big-bud" distortion of a currant bud by tarsonemid blackcurrant gall mites (*Cecidophyopsis ribis*) feeding.

Figure 3.16   An adult mottled woodlouse (*Oniscus asellus*).

Figure 3.17   Adult and immature woodlice overwintering under a brick.

gray and other species may be more colorful, including a few that are a rose-pink color. Adult sizes range from about 2.5 up to 18 mm in length. They all live in damp habitats such as under decaying wood and plant matter, stones, and rocks (Figure 3.17). Rockeries are particularly favorable because of the moist organic matter interspaced with stones. Indeed, a moist environment is required for their survival, as they will quickly die if exposed to warm, dry air.

Although regarded as pests when they feed on seedlings, fruit, and young plants, they also feed on dead plants, animal remains, fungi, and dung, thus performing a valuable role in recycling nutrients. On many plants, damage resembles slug feeding but without the characteristic slime trail. Old feeding areas may become heavily scarred with elongated calluses. The feeding site depends upon the species: Some feed below soil level (*Porcellio* spp.), and others girdle the base of stems (*Armadillidium* spp.) (especially in organically grown cucumbers); some eat fruit and leaves, particularly in very humid environments (*Oniscus* spp.).

The segmented body has a clearly defined head region with a pair of compound eyes, two pairs of antennae (one pair is minute and hardly discernible), a pair of mandibles armed with teeth, and small, jointed appendages used in the manipulation of food. The seven-segmented thoracic region makes up the majority of the body and carries a pair of simple jointed legs on each segment. A narrow, six-segmented abdomen often ends in a tail and, in some species, the abdomen forms a cavity containing a moisture-rich pouch within which there are air tubules that act rather like a gill, allowing the animal to "breathe." Gravid females have a brood pouch formed by pairs of overlapping plates growing from the legs of thoracic segments two to five. Depending on species, up to 200 eggs (commonly 20–50) may be carried in the pouch for up to 1 month before hatching. Hatchlings are white, about 1 mm in length, and they remain in the brood pouch for a week or so before dispersing. They attain their adult coloration after about a fortnight and molt several times before reaching adulthood.

Biocontrol

The principal natural enemies of woodlice include centipedes, particularly lithobiids, which can kill many young woodlice. Several ground beetles, including staphylinids and carabids, are recorded as feeding on woodlice. Parasitoid wasps, often specific to certain species of woodlice, can be found worldwide. Spiders of the Clubionid and Dyseridae families are reported to include woodlice in their diet. Other predators of woodlice include birds, reptiles, and mammals. Probably the easiest way of controlling or reducing woodlice is to remove additional sources of food and all hiding places and to dry the area as much as possible.

Millipedes

Commonly known as thousand legged worms (Figures 3.18 and 3.19), these round bodied arthropods have four legs on each segment and can vary in size from a few millimeters to over 20 cm in length. They are usually restricted to moist places where they feed on organic matter, either plant debris or live plant

Figure 3.18 A flat-backed millipede adult (*Polydesmus angustus*) soil pest.

Figure 3.19 A flat-backed millipede (*Polymicrodon polydesmoides*) soil pest.

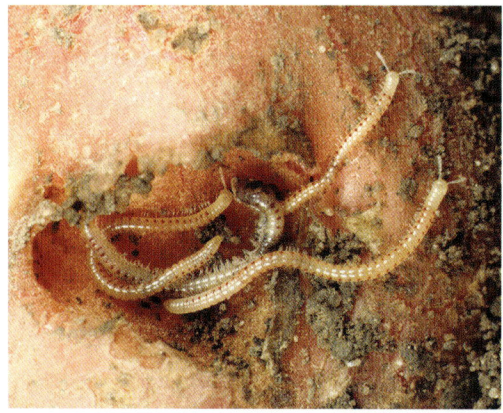

Figure 3.20 Spotted snake millipedes (*Blaniulus guttutalus*) on a damaged potato tuber.

tissue such as roots and tubers (Figure 3.20), low lying leaves, and stems. If handled roughly, most species produce a foul smelling odor that acts as a defense against predatory attack.

Biocontrol
Some species of nematodes and parasitoid wasps in the family Sciomyzidae are reported as attacking millipedes, but there are no commercially available biocontrol organisms.

Leafhoppers, Psyllids, Whitefly, Aphids, Scale insects, and Mealybugs

All of these insects possess piercing mouth parts made up of needle-like mandibles that penetrate the outer layer of plant cells. The mandibles and maxillae probe deeply into the tissues and, in doing so, transfer some of their saliva in order to begin the digestion process. This saliva may carry plant viruses, which infect the plant causing disease, or the saliva may cause leaf distortion (as with the common green capsid bug for which there is no commercial biological control).

These pests obtain amino acids from the plant sap that they use for growth, but as the concentration is relatively low, considerable quantities of sap are needed. However, the sap is rich in sugars and a large amount of a sugary waste product, called honeydew, is deposited on the leaves. This is a valuable food source for many other organisms. A fungus, *Cladosporium* sp., frequently grows on this honeydew, resulting in a black sooty mold on the plant surface (Figure 3.21) that excludes light from reaching the leaves

Figure 3.21 Sooty mold (*Cladosporum* sp.) growing on aphid honeydew on rose leaves; note empty cast skins of aphids stuck in sticky honeydew.

and can severely interrupt photosynthesis. Plant damage and serious economic losses can occur in this way.

This group of pests is commonly found throughout the world. They frequently form the diet of many parasitoid and predatory insects. Several pathogenic fungi and some mites also kill these insects. Some of the parasitoids are specific to just a few species while, in general, predators will feed on many species and even other insects.

LEAFHOPPERS

Adults and nymphs resemble elongated aphids (Figure 3.22), but can be much more mobile, particularly in the later life stages. Adults are easily disturbed and have a short flight, so they appear to jump between leaves and plants. Adult size varies from 2.5 to 10 mm depending on species. They can be very colorful with spots, stripes, and dapples aiding their camouflage among leaves. Feeding is by the insertion of a proboscis that sucks sap. Air enters the empty plant cells and gives rise to the characteristic symptom. These vary with leaf type; soft leaves are most severely affected, showing short yellow/white chains (Figure 3.23) leading to almost completely bleached leaves. Such symptoms can occur on chrysanthemum, salvia, and tomato. Leaves of rhododendron and fruit trees show less visible damage except where very high numbers of leafhoppers occur.

Females lay up to 50 eggs singly into veins on the undersides of leaves. These hatch after 5–10 days to minute wingless nymphs, which can take up to 2 months to reach adulthood. The common name "ghost fly" has been given to leafhoppers due to the rapid appearance of damage and only transparent skins left on the leaf (Figure 3.24). These are the cast skins of nymphs, still attached to the leaf by the insect's proboscis. Feeding on young seedlings can lead to plant death while damage to older plants is mainly cosmetic. Several species can transmit viruses while *Graphocephala* spp. are associated with rhododendron bud blast disease.

Biocontrol
The egg parasitoid *Anagrus* spp. is commercially available in some countries, but leafhoppers are usually controlled by predators such as lacewings (*Chrysoperla* spp.), ladybirds (*Coccinella* spp.), and anthocorid bugs.

Figure 3.23   Leafhopper damage to basil leaves.

Figure 3.22   A sage leafhopper (*Eupteryx melissae*) feeds on herbs and chrysanthemum.

Figure 3.24   A rose leafhopper (*Edwardsiana rosae*) adult (facing empty cast skin) with nymphs on a rose leaf.

PSYLLIDS

Psyllids or suckers are small, aphid-like insects (Figure 3.25); adults have large, membranous wings but they tend to jump rather than fly. Most species are free living, although some form galls in which eggs are laid. Trees such as alder, apple, ash, birch, box, eucalyptus, hawthorn, and pear form the main host plant range. Psyllids tend to be host-specific pests and can cause considerable damage when they attack growing stems and flowers. Nymphs (Figure 3.26) move slowly and are flattened and, in some species, can resemble scale insects. They produce a white, waxy secretion that covers and protects them. Copious honeydew is produced, most of which remains in wax-covered droplets that prevent the insects from becoming sticky.

Biocontrol

Psyllids have better natural protection than aphids, but are attacked by anthocorids, lacewings, ladybirds, and other generalist predators.

WHITEFLIES

Various species are common on a range of protected and field vegetable crops throughout the world. In cotton and field-grown tomato crops, whole clouds of adult whitefly, usually a *Bemisia* strain, may be disturbed and found on the wing at one time. The *Bemisia* genus (Figure 3.27) is probably the most economically damaging and is subdivided into a number of new species as well as various strains (e.g., *Bemisia tabaci* A, B, and so on). These divisions reflect the nature of the crop damage inflicted on plants. *Bemisia argentifolii*, the silver leaf whitefly, produces a leaf silvering symptom.

Whiteflies have a number of serious effects on plants. Their presence alone detracts from the plant value (Figure 3.28). In addition they secrete copious quantities of honeydew on which a sooty mold grows. They are also vectors of some very important plant viruses, including tomato yellow leaf curl.

Pesticide resistance is frequent in many populations; however, some of the new specific insecticides give better control and, with current pesticide resistance strategies, these should remain active for several years.

Figure 3.25   **An adult box sucker (*Psylla buxi*) on box leaf (*Buxus* spp.).**

Figure 3.26   **An apple sucker nymph (*Psylla mali*) on an apple leaf.**

Figure 3.27   **An adult tobacco or sweet potato whitefly (*Bemisia tabaci*) with freshly deposited eggs.**

Figure 3.28   Adult glasshouse whitefly (*Trialeurodes vaporariorum*).

Figure 3.31   A *Trialeurodes vaporariorum* "pupa" showing a red eye inside raised sides and waxy filaments on surface.

Figure 3.29   An adult female cabbage whitefly (*Aleyrodes proletella*) with a circle of eggs on a white waxy surface.

Figure 3.32   A pair of adult female cabbage whitefly (*Aleyrodes proletella*) with a smaller male among emergent "pupae."

Figure 3.30   Third instar larval "scales" of *Bemisia tabaci*.

Adults are small (1.25–2 mm), gray- to white-winged flies. Some species have bands or spots on their wings and they are usually found at rest on the underside of leaves (Figure 3.29). Eggs are initially white and laid singly or in complete or partial circles. After a few days, they melanize to an almost black color before hatching to a six-legged nymph. This "crawler" stage moves a short distance to locate a suitable site for larval development. The four larval instars are commonly known as whitefly scales and may take between 7 and 20 days to develop (Figure 3.30). The last instar is referred to as a pupal stage, although, as no molt occurs, it is a false pupal stage (Figures 3.31 and 3.32). In some species the "pupa" may be covered in a waxy secretion making it difficult to see clearly (Figures 3.33 and 3.34). Adults emerge some 10 days later and start to feed almost immediately; oviposition usually begins 1–2 days later. Adult whiteflies may

Figure 3.33   Scale-like "pupa" of viburnum whitefly (*Aleurotrachelus jelinekii*).

Figure 3.35   A rose-grain aphid female (*Metapolophium dirhodum*) with offspring on a wheat leaf.

Figure 3.34   Viburnum whitefly "pupae" (*Aleurotrachelus jelinekii*) showing white waxy growths.

Figure 3.36   The pink form of rose aphids (*Macrosiphum rosae*) on a rose leaf.

live for several weeks. The common glasshouse whitefly (*Trialeurodes vaporariorum*) has been widely studied as a pest for over 100 years and numerous commercial biocontrols are available. Other species, such as the cabbage whitefly (*Aleyrodes proletella*) (Figure 3.29), may be attacked by several beneficials; although they can be encouraged, none are commercially available.

Biocontrol
Whiteflies may be parasitized by specialized wasps, predated on by heteropteran bugs and coccinellids or infected by several entomopathogenic fungi. Yellow traps are useful in reducing high numbers of adults but may also catch natural enemies.

APHIDS
Commonly known as blackfly, greenfly, greenbug, and plant lice, aphids are small,

Figure 3.37   Vetch aphids (*Megoura viciae*) on the stem of a broad bean plant.

soft-bodied insects ranging in size from 0.5 to about 6.5 mm in length. Colors can vary between different species and even between strains of the same species, ranging from shades of green through to black, yellow, chestnut brown, pink, gray, and waxy white (Figures 3.35–3.37). Several species have

Figure 3.38   A colony of black-bean aphids (*Aphis fabae*).

Figure 3.39   Symptoms of barley yellow dwarf virus (BYDV) transmitted by bird cherry aphid (*Rhopalosiphum padi*).

darker markings on a lighter toned body or vice versa (Figure 3.38), camouflaging them against their plant background but aiding identification. To identify a species fully it may be necessary to check the frontal outline of the head between antennae and associated tubercules, the abdominal siphunculi (exhaust pipes at rear), and rear cauda (flicking tail). Aphids may be differentiated from other Hemiptera by the ability to produce active young from unimpregnated females (parthenogenesis). The same species may be present as alate (winged) and apterae (wingless) individuals simultaneously on the same leaf. Alate aphids are frequently produced in response to adverse conditions such as overcrowding, poor plant condition, the onset of winter conditions, or for migratory purposes.

The biological life cycle of aphids can be one of three types:

- The monoecious holocycle is the simplest and most common on outdoor crops; all generations develop on the same host plant species. Overwintered eggs hatch into parthenogenetic females in the spring. These produce both alate and apterous aphids through the summer months and a sexual generation in the autumn that, after mating, lay eggs to survive the winter.

- With the dioecious holocycle, the aphid has two different host plants: a woody winter host in which eggs are laid to overwinter and a nonwoody summer host that supports the bulk of generations. Usually, two generations occur after the eggs hatch in the spring, followed by a migration to the main summer host plant on which multiple generations (both winged and wingless) are produced. As discussed previously, a sexual generation develops in the autumn that migrates back to the winter plant for egg production.

- Anholocyclic aphids are alate or apterous parthenogenetic females, depending on plant condition and colony density. This type is common in the Mediterranean, subtropical, and tropical regions where winter conditions are very mild to nonexistent. They also occur on many protected crops.

Aphids feed by penetrating plant tissue with four ultrathin stylets held within a protective rostrum that, at rest, is folded against the underside of the insect. Only the stylets enter the plant as the insect feeds, injecting saliva by way of a small canal formed between the hypodermic-like stylets and sucking up the partially digested sap. During feeding, aphids (and many other sap-feeding insects) can transmit plant viruses (Figure 3.39), cause galls to form, and produce leaf blistering, as with *Cryptomyzus ribis* on black currant (Figure 3.40), or leaf curling, as with *Brachycaudis helichrysi* on peach and plum. Large volumes of sap are taken in during

Figure 3.40   Currant aphid (*Cryptomyzus ribis*) feeding on black currant causes damage in the form of raised pink leaf blisters.

Figure 3.41   Brown scale insects (*Parthenolecanium corni*) on the stem of a grapevine.

Figure 3.42   Soft brown scale insects (*Coccus hesperidum*) with honeydew on the leaf surface.

feeding and almost equal amounts of sugary water excreted as honeydew. This is flicked away from the aphid by the cauda, to fall on the upper surface of leaves and stems. High numbers of aphids can give rise to copious quantities of sticky honeydew, which in turn supports the growth of a fungus turning the whole area black (see Figure 3.21). Ants are frequently associated with aphids and it is often assumed that they are feeding on the insect when they are in fact feeding on the sugary honeydew. They also commonly defend aphids against attack from parasitism and predators (see Figure 4.147 in Chapter 4). Several aphid species live below the ground; feeding on plant roots, root aphids may go unnoticed until the plants begin to wilt.

Biocontrol
Aphids are very common and numerous and are a major source of nutrition for organisms used in biological control. Being one of the earliest pest organisms to establish on plants, they are parasitized by several hymenopteran wasps, predated on by almost every group of insect and mite predator, and attacked by several fungal pathogens.

SCALE INSECTS
Adults range in size from 3 to 7 mm and can be circular or elongated. All have a waxy skin that protects the body (Figure 3.41). Soft scale insects (Coccidae) produce honeydew (Figure 3.42), often in large quantities

giving rise to sooty mold. Armored or hard scale insects (Diaspididae) do not produce honeydew. Scale insects are hardier than mealybugs and have a larger geographical range. They are found on several outdoor plants throughout Europe and North America (especially on crops of citrus), but are most frequently associated with conservatory plants. As with mealybugs, males are winged and can make up to 50% of the population. Some species produce several hundred eggs

Figure 3.43   Brown scale insect female (*Parthenolecanium corni*) turned over to show young nymphs developing under her body.

Figure 3.44   Cottony cushion scale insects (*Icerya purchasi*).

(Figure 3.43), while others are viviparous. Young nymphs are known as crawlers and can move several centimeters from their place of birth before settling for the rest of their development.

The body of young, hard scale insects is a delicate membrane structure that settles on the plant and molts, losing all its legs and becoming immobile for the rest of its development. For protection it binds its fecal material with a wax secretion to form a hard covering shell (Figure 3.44). The mouthparts of hard scale insects are extremely long and housed in a pouch inside the body. When extended, they can feed almost anywhere. Generally, they are not phloem feeders, so little or no honeydew is produced. The mouthparts can penetrate the plant tissue to feed on cells at a distance of several times the insect's diameter. Eggs and nymphs can be transported on wind currents, in water droplets, by insects and birds, and frequently by man when transporting plants around the world. Plant damage can be seen as yellowing foliage, weakened or stunted growth, sunken spots that correspond to a mature scale insect on the underside of leaves (as can be found on banana), and the growth of sooty mold from honeydew residues.

Biocontrol
Several host-specific parasitoids are associated with scale insects. Various ladybirds are predatory against scale insects including

Figure 3.45   A glasshouse mealybug (*Pseudococcus viburnii*) colony on an Easter cactus *Epiphyllum*.

Chilocorus bipustulatus and *Rhodiola cardinalis*. The Australian mealybug destroyer, *Cryptolaemus montrouzieri*, will also feed on these pests, as will lacewing larvae and other generalist predators. Parasitoid nematodes used against sciarid fly larvae have given excellent control when sprayed onto leaves (the leaf surface must remain wet for several hours after application for best results).

MEALYBUGS
Of tropical and subtropical origin, these sap-sucking insects are now found on protected plants throughout the world, particularly those that are slow growing and permanently sited such as in interior atriums. The species *Pseudococcus viburnii* produces large white masses of eggs at the rear (Figure 3.45) and can be a major pest on glasshouse pepper and tomato crops. This problem has become more

Figure 3.46   A citrus mealybug (*Planococcus citri*); note the dark bar along the back.

Figure 3.48   A citrus mealybug (*Planococcus citri*) with orange eggs in a protective wax "cotton wool."

Figure 3.47   Long-tailed mealybugs (*Pseudococcus longispinus*) on a palm leaf.

severe since the length of the cropping cycle of both tomatoes and peppers is now frequently 48 to 50 weeks with a 2- to 4-week turnaround before the next crop is planted. They are also common on many plant collections. In warmer regions, mealybugs can be found on citrus and other orchard crops where severe economic damage may be caused. Several viruses are also transmitted by mealybugs; the most important is cacao swollen shoot virus.

Their bodies are covered with dusty white wax filaments, some of which form a fringe around the insect (Figure 3.46) and may be used to help identify the various species. Citrus mealybug (*Planococcus citri*) has a characteristic dark band with no wax down the middle of its back and no obvious waxy terminal filaments. The long-tailed mealybug (Figure 3.47)

(*Pseudococcus longispinus*) has a dark band similar to that of the citrus mealybug with a whiter body and wax tail filaments as long as the body. The long-tail mealybug gives birth to live young that initially congregate close to the parent. Species such as *P. citri* and *P. viburnii* produce egg clusters covered in a protective wax deposit (Figure 3.48) that may be found at the rear of the body.

They are sap feeders. Generally the young crawler stage has short mouthparts and feeds on the phloem of the leaf. They migrate to the stem as they mature and their now longer mouthparts can reach the vascular bundles. In high numbers, stem girdling can kill branches or even whole plants. Several species of mealybug produce copious quantities of honeydew that encourages black sooty mold. Toxins can be injected as mealybugs feed, causing distortion and yellowing of foliage in some plants. Although mobile, they tend to remain localized, frequently moving slowly along stems as the plant grows.

Clumps of mealybugs can become too heavy and fall to lower leaves or the ground, where they disperse to start new colonies. Male mealybugs are delicately winged and can make up to 50% of the population. Males mature early and after repeatedly mating they die within 48 hours. In most common species sexual reproduction is the norm, with females of some species giving birth to live young while others produce masses of eggs covered in a protective white woolly wax. Unmated female mealybugs can survive away from plants for several months, often hidden in inaccessible places such as behind plant ties, within hollow support canes, under pot rims, and among leaf debris.

Biocontrol

The Australian ladybird, *Cryptolaemus montrouzieri,* is commercially used throughout most of the world and provides adequate control of several species. Specific parasitoids are becoming more widely available and can be used in association with predators. Other generalist predators such as lacewings are also found feeding on young nymphs. Predatory mites (*Hypoaspis* spp.) feed on root mealybugs in compost and foliage species when they are close to the ground. Entomopathogenic fungi provide good control as long as high enough humidities can be maintained to enable infection.

THRIPS

Adults are tiny, winged insects ranging in color from light golden yellow to brown (with and without stripes) to black with and without stripes (Figure 3.49). They have several common names, including thunder bugs (Figure 3.50), but may also be named after the plant or geographical region from which they originated. For instance, gladiolus thrips (*Thrips simplex*) is a pest of gladioli flowers

(Figure 3.51) and Western flower thrips (*Frankliniella occidentalis*) is a pest that originated on the west coast of North America, although it is now found throughout the world (Figure 3.52). The poinsettia or impatiens thrips (*Echinothrips americanus*) (Figure 3.53)

Figure 3.50   An adult female cereal thrips (*Limothrips cerealium*).

Figure 3.51   Gladiolus thrips larvae (*Thrips simplex*).

Figure 3.49   A rose thrips (*Thrips fuscipennis*) on a rose petal.

Figure 3.52   An adult female Western flower thrips (*Frankliniella occidentalis*).

COMMON PEST SPECIES

Figure 3.53   A colony of black thrips (*Echinothrips americanus*); this is a gregarious species with adults and larvae living in close proximity to each other.

Figure 3.54   Thrips feeding damage on deeply colored flowers such as this *Dahlia* shows as fine, silver-white streaks.

Figure 3.55   Thrips feeding damage on chrysanthemum leaf shows as a fine "silvering" on leaves; close inspection with a hand lens would reveal small black frass pellets.

Figure 3.56   Tomato spotted wilt virus (TSWV) on cyclamen shows as concentric ring spots.

is a major pest on many protected ornamental plants. It produces a repellent secretion that deters most predators from attacking it. Thrips damage leaves, fruit, and flowers (Figure 3.54) by piercing and sucking out the cells of the surface membranes, leaving silvery/gray patches littered with small black fecal pellets (frass) (Figure 3.55). Leaf and fruit distortion can occur when thrips feed on very young developing parts of the plant. This can cause severe economic losses. Several thrips species are also responsible for transmission of plant viruses such as tomato spotted wilt (Figures 3.56 and 3.57). This disease causes severe economic losses in a number of crop plants. In some countries and states, certain species of thrips are notifiable quarantine pests and their presence may lead to crop destruction or, more usually, a chemical spray program.

Adult females oviposit into plant tissue and may leave a small raised blister where the egg was laid. Eggs hatch after 2–12 days into minute first instar larvae, which last only a few days before molting to the second instar

Figure 3.57   Spotted wilt virus on alstromeria showing leaf discoloration and distortion.

Figure 3.58   A tobacco budworm moth (*Helicoverpa virescens*) on a cotton leaf.

larva (4–10 days). These develop to prepupa and pupa. Pupal stages of thrips may occur on the plant or in the growing media and can last for a few days to several months depending on temperatures. Adults have a preoviposition period of 2–10 days before they commence their next generation.

Biocontrol
Predatory mites (*Hypoaspis* spp.) can kill thrips pupae if they pupate on the ground. Parasitoid nematodes, parasitoid wasps, predatory Hymenoptera, and entomopathogenic fungi are commercially used to control larvae and adults. The most widely used control agents are various *Amblyseius* mites. These mainly attack the first instar larva and are routinely used as a preventive measure.

Figure 3.59   Caterpillars of diamondback moth (*Plutella xylostella*) feeding on a cabbage leaf.

Lepidoptera
CATERPILLARS (OF MOTHS AND BUTTERFLIES)
Lepidopterous caterpillars are the larval stages of moths and butterflies (Figures 3.58–3.61). In almost all species they are plant feeders equipped with powerful jaws capable of eating through most plant tissue, including roots, young stems, leaves, flowers, and fruit (Figures 3.62–3.64). Damage may be seen as complete defoliation of a particularly favored host plant or, more subtly, as a weakening of plants by root feeders and stem borers. Tortricid caterpillars protect themselves by

Figure 3.60   An adult diamondback moth (*Plutella xylostella*).

feeding within rolled leaves or growing tips that are stitched together with a silken thread, while others from this family, such as the codling moth (*Cydia pomonella*), which is found in apples and cutworm in turnips, feed inside fruit

Figure 3.61   An adult cabbage white butterfly (*Pieris brassicae*).

Figure 3.63   A tomato moth (*Laconobia olreracea*) feeding on tomato fruit.

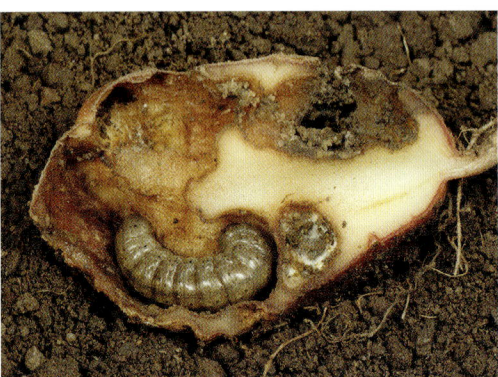

Figure 3.64   A common or turnip cutworm caterpillar (*Agrotis segetum*) inside a damaged potato tuber.

Figure 3.62   A cotton bollworm, corn earworm, or Old World (African) bollworm caterpillar (*Helicoverpa armigera*) feeding on a corn cob.

Figure 3.65   Apple leaf miner caterpillar (*Leucoptera scitella*) mines in an apple leaf with black lines of frass deposits.

or roots. Many Lepidoptera feed inside leaves and may be mistaken for leaf mining flies (Figure 3.65), but they can easily be identified by the production of silk from spinnerets found behind the head. This silk forms a protective cover and allows them to escape attack by parasitoids or predators. Other species produce copious quantities of silk that make protective tents in which they develop (Figure 3.66).

Figure 3.66   Small eggar moth caterpillars (*Eriogaster lanestris*) on tent built in an oak tree.

Figure 3.67   Neonate larvae of a cabbage white butterfly (*Pieris brassicae*) with their empty egg cases.

Figure 3.68   Silver Y moth caterpillar (*Autographa gamma*) feeding on sage leaf; note two pairs of prolegs.

Figure 3.69   A tomato moth caterpillar (*Laconobia oleracea*) feeding on tomato foliage; note four pairs of prolegs.

Figure 3.70   Mountain ash sawfly larva (*Pristiphora geniculata*) feeding on *Sorbus*; note six pairs of prolegs.

Eggs may be laid singly or in small groups; these hatch to minute neonate larvae (caterpillars) that initially eat their own egg cases (Figure 3.67) before feeding on plant tissue. The generally undifferentiated thorax (segments immediately behind the head) may bear a hard protective plate that has three pairs of true legs (except in some of the leaf mining caterpillars, where the first instar is often legless). The abdomen has two to four pairs of fleshy prolegs (Figures 3.68 and 3.69), used for walking as well as clinging onto surfaces, and the rear of the larva has an anal plate and a terminal clasper with crotchets. Sawfly (order Hymenoptera) caterpillar may look similar to Lepidoptera larva, but it has five or more pairs of prolegs (Figures 3.70 and 3.71). Identification to insect order is important if using *Bacillus thuringiensis* as a control treatment against caterpillar.

Figure 3.71   Gooseberry sawfly (*Nematus ribesii*) feeding on gooseberry leaves; note seven pairs of prolegs.

Figure 3.72   Gregarious buff tip moth caterpillars (*Phalera bucephala*) on damaged oak foliage.

Figure 3.73   A cluster of buff tip moth caterpillars (*Phalera bucephala*) on damaged oak foliage.

Caterpillar skin can be smooth, bristly (Figures 3.72 and 3.73), or very hairy. In some species, the hairs are irritating and act as a defense mechanism when the insect is attacked. Color and body form can vary dramatically, making both caterpillars and adult moths or butterflies extremely varied. Developing larvae must periodically shed their skins as they grow and, once fully developed, form a pupa. In some species this is the over-wintering stage and the pupa may remain hidden for several months before the adult moth or butterfly emerges.

Biocontrol
Many predators and parasitoids attack and kill caterpillars, including parasitoid wasps (Hymenoptera) and parasitoid flies (Diptera), which lay their eggs inside young larvae. The minute *Trichogramma* spp. lay their eggs into butterfly and moth eggs. Predators include bugs (Hemiptera) such as Anthocoridae and lacewing larvae (*Neuroptera* spp.). Birds and beetles will also devour caterpillars and pupae. Several pathogens attack Lepidoptera, including bacteria; *Bacillus thuringiensis* is the most widely found and commercially used form of biological control agent. Fungi and virus pathogens are also common and are beginning to be used commercially as spray treatments. They are host insect-specific, usually affecting and killing the larval stage. Pathogenic nematodes are also commercially available as either crop sprays or as a drench treatment against those with a soil stage such as the swift moth (*Hepialus lupulinus*).

Flies
This very large group includes true flies or two-winged flies, which have one pair of wings and the remnants of hind wings that are reduced to a pair of club-shaped balancers or halteres. Their form can vary from slender mosquitoes to stout bodied houseflies and hoverflies. Color can also range from the drab grays and browns of bean seed fly, (Figure 3.74) and carrot and cabbage rootflies to the bright metallic hues of blue and green blowflies or bottleflies (*Calliphora vomitoria* and *Lucilia illustris,* respectively). Several, such as hoverflies (Figures 3.75 and 3.76), mimic other flies and wasps by having

Figure 3.74   Adult pea and bean seed fly (*Delia platura*) on pea leaf.

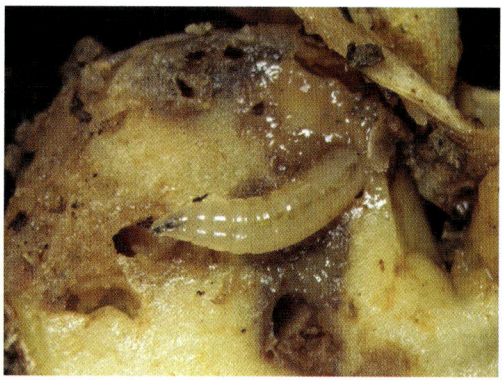

Figure 3.77   Pea and bean seed fly larva (*Delia platura*) feeding on pea seedling.

Figure 3.75   Adult large narcissus fly (*Merodon equestris*) is a pest species of hoverfly.

Figure 3.76   Large narcissus fly larva (*Merodon equestris*) in sectioned daffodil bulb.

Figure 3.78   Pea and bean seed fly (*Delia platura*) damage to bean seedlings.

yellow and black stripes. They are all liquid feeders with sucking mouthparts but frequently possess piercing organs with which to penetrate their target source of food. This can include blood from animals, body fluids from other insects, and, commonly, plant tissue.

The larval stages have no legs, although they may have bristles or short, fleshy stumps similar to the prolegs of caterpillars. Feeding is usually by a pair of mouth hooks (Figures 3.77 and 3.78). The following sections detail just a couple of examples of flies that are common plant pests.

Figure 3.79   A South American leaf miner larva (*Liriomyza huidobrensis*) exposed from a mine in a tomato leaf.

Figure 3.80   South American leaf miner pupa (*Liriomyza huidobrensis*) on a tomato leaf with adult fly puncture marks. *Liriomyza* species of leaf miners pupate outside the leaf surface.

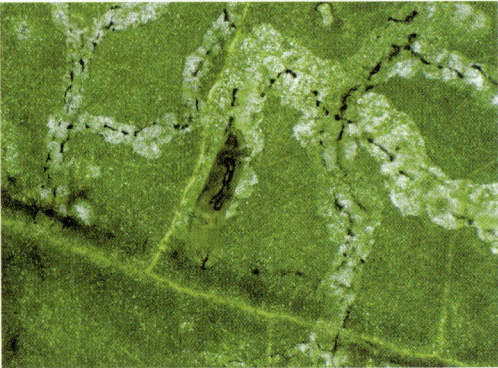

Figure 3.81   An American serpentine leaf miner larva (*Liriomyza trifolii*) tunneling in a mine in a tomato leaf. The continuous black frass trail indicates a *Liriomyza* spp.

Figure 3.82   A chrysanthemum leaf miner (*Chromatomyia syngenesiae*) trailing mine in common or smooth sow thistle (*Sonchus oleraceus*) with larva at the tunnel end and showing discontinuous frass trail along length of the mine.

LEAF MINERS

Larvae mine in the central tissues of the leaf, sometimes nearer to the upper surface (Figure 3.79) and sometimes nearer the lower one. They never mine near to both surfaces. In severe infestations, they can tunnel in flowers and plant stems, killing developing growth. Apart from disfiguring ornamental plants, high numbers of mines can lead to economic loss of vegetable and salad crops. Adult females (2.5–3 mm long) puncture leaves with their ovipositor, leaving open pits or puncture holes (Figure 3.80 and 3.81) from which both sexes feed on plant juices. In some pits an egg is deposited that hatches into a minute larva and begins tunneling in the leaf, forming a characteristic mine (Figure 3.82). Some species create a spiral or

serpentine tunnel, while others produce mines that follow the leaf veins. The mine shape and frass trail within can help to differentiate the species of leaf miner.

Mature larvae (3–3.5 mm long) can either pupate in the leaf as a small blister—usually on the underside of mined leaves (Figure 3.83)—or, depending on species, drop out of the leaf to produce a puparium. Those that pupate externally may land on the upper surface of lower leaves or drop through to the soil, where they may remain for several months while they overwinter. Leaf miner larvae may be parasitized by specialized wasps, predated

Figure 3.83 *Chromatomyia syngenesiae* pupate under the leaf surface and emerge through the leaf epidermis.

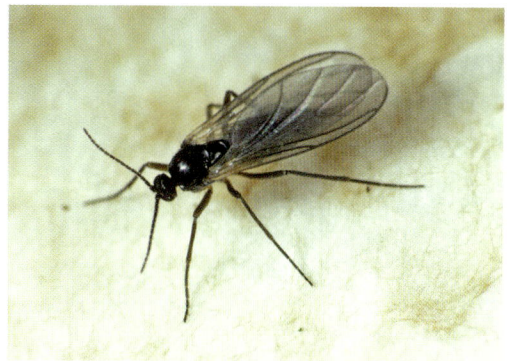

Figure 3.85 Adult sciarid fly (fungus gnat) (*Lycoriella auripila*); the glasshouse sciarid fly (*Bradysia paupera*) looks similar to the mushroom sciarid fly.

Figure 3.84 An American serpentine leaf miner (*Liriomyza trifolii*) adult fly.

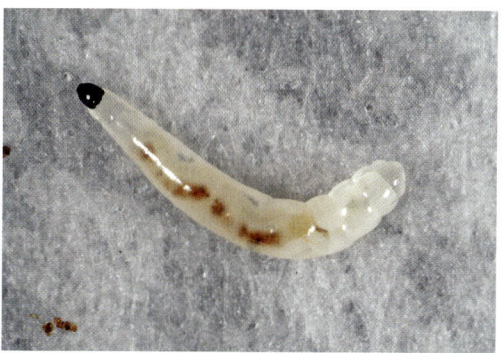

Figure 3.86 A sciarid fly larva (*Bradysia paupera*).

on by heteropteran bugs, or eaten as pupae by ground beetles. Adult leaf miner flies (Figure 3.84) can be caught on yellow sticky traps, with horizontally hung traps catching considerably more than vertical ones.

SCIARIDS

Commonly known as fungus gnats, these narrow black flies (Figure 3.85), which are 2.5 mm long run over compost and deposit eggs just below the surface. *Bradysia* spp. are commonly found in glasshouses, whereas *Lycoriella* spp. are common in mushroom crops. The larvae are almost transparent, with a shiny black head (Figure 3.86), and can reach up to 6 mm in length before pupating close to the compost surface. Black and white threads inside the larva are the gut and fat deposits. Larvae feed on young root tissue, stunting the

plant growth. Cuttings and young cucumber plants can be tunneled, often leading to their death. Both adults and larvae can exacerbate fungal root pathogens such as *Pythium* and *Thielaviopsis* spp. Adults may carry spores of *Pythium*, especially in greenhouses and *Verticillium* sp. in mushroom crops. Damage to roots of young plants can create an entry point for *Theilaviopsis*. Moist composts with high organic matter content are most favored, although it is not uncommon to find them in artificial media such as stone wool or even hydroponic growing systems.

Control of sciarid larvae can be achieved by predatory mites (*Hypoaspis* spp.), parasitoid nematodes, parasitoid wasps, pathogenic bacteria (specific strains of *Bacillus thuringiensis*), or entomopathogenic fungi (*Furiae sciarae*). The latter are naturally occurring and a dead larva resembles a thread of white cotton on the compost surface.

SHORE FLIES

Many growers mistake fungus gnats for shore flies (*Scatella stagnalis* and *S. tenuicosta*), which are in a different family: Ephydridae. Adult flies are slightly longer (3 mm) and broader (1.25 mm wide) (Figure 3.87). When at rest on a plant, compost, or other flat surface, their folded wings appear to have three or four almost transparent white spots on them. Shore flies feed mainly on algae, but the larvae may cause slight damage to roots. Adults sitting on leaves may leave black fecal spots that, on edible crops such as herbs, affect quality and sometimes cause health concerns. The larvae are white and segmented with a small black head. They feed in and on algae that may be growing on almost any surface, such as compost, benches, paths, and floors. A hard-skinned puparium develops in the algae-covered compost surface and, depending on temperatures, the adult may emerge after only 3–4 days, giving rise to rapidly increasing populations. Control is best achieved by eliminating their algal food source, although biological controls as for sciarids (such as *Atheta coriaria*, *Coenosia attenuata*, and some entomopathogenic nematodes) can feed on *Scatella* spp.

Beetles

This is the largest order of insects, with over a million known species found throughout the world; some are able to survive in extreme environments. This section will deal with some of the plant-damaging species. Many are predators and several of these are included in Chapter 4. Vine weevil (*Otiorhynchus sulcatus*) is a common and serious beetle pest and is described separately (see p. 62).

All beetles have biting mouthparts capable of chewing into hard surfaces, such as seed grains (Figure 3.88), leaves (Figure 3.89), insect bodies, roots, stems, and live or dead wood. In most species, the larva is most damaging (Figures 3.90 and 3.91). It, too, has biting jaws and usually feeds on similar food to the adult.

Some of the more common pest species living in soil and feeding on plant roots are in the group Scarabaeidae which include

Figure 3.88   Saw-toothed grain beetle (*Oryzaephilus mercator*) feeding on barley.

Figure 3.87   An adult shore fly (*Scatella stagnalis*).

Figure 3.89   An adult female lily beetle (*Lilioceris lilii*) with her eggs on a lily leaf.

Figure 3.90   Larvae of *Lilioceris lilii* cover themselves in frass as a defense against predation. Feeding damage initially shows as "windows" where the grubs have eaten the leaf cells; these open to leave a shredded appearance.

Figure 3.92   A cockchafer grub larva (*Melolontha melolontha*).

Figure 3.91   Viburnum beetle larvae (*Pyrrhalta viburni*) and damage to viburnum foliage.

Figure 3.93   An adult common cockchafer, also known as May or June bug (*Melolontha melolontha*).

chafer grubs (Figure 3.92) and chafer beetles (Figure 3.93). These have a life cycle spanning 1–3 or more years. Adults are usually found from late spring to early summer and once mated, the females burrow into friable, free draining soil to lay eggs which soon hatch into larvae that feed mainly on grass roots or on the roots of lettuce, strawberry or several ornamental shrubs. Larvae feed to a depth of 5–15 cm until late autumn, when they burrow deeper and produce an earthen cell in which they overwinter. Pupation occurs within the cell during spring to produce fresh adults the following season. Predatory ground beetles feed on the eggs and larvae and provide reasonable control but in areas where natural enemies have been depleted, such as golf courses, heavy pesticide use may be necessary. More environmentally friendly

entomopathogenic nematodes, such as those used for vine weevil control, have been successfully used against a range of soil dwelling pests.

Beetles from the family Elateridae are known as click beetles or skip-jacks after their ability to flick themselves into the air with an audible click when they fall on their backs. The larvae are known as wireworms (Figure 3.94) and can cause considerable damage to many plant roots and tubers, particularly in the garden but also the roots of grasses and, more rarely, agricultural crops. Permanent grasslands contain the greatest populations of click beetles, and the most damage is done when these areas are used for arable crops or unsterilized turf is used as a basis for a soil compost or to make up beds in glasshouses. The larvae possess six almost invisible legs and have a firm cylindrical body, covered with a tough skin, and a dark brown head with powerful biting mouthparts to eat into many host plant roots and stems. They may spend 4–5 years as larvae before developing to a pupa and then adult in the following spring. The principal beneficial organisms that attack wireworms include birds, several predatory beetle larvae, and pathogenic fungi.

The family of beetles known as the Chrysomelidae are mostly leaf feeders and include the tiny flea beetle that bores small holes in leaves (Figure 3.95). Most flea beetles are very host plant specific and, when a particularly favored plant is grown on a commercial scale, severe economic losses can occur (Figure 3.96). Due to their voracious appetite and potential narrow host range, flea beetles are successfully used for the biological control of several weed species in Canada and the United States.

The common asparagus beetle (*Crioceris asparagi*) is an important pest throughout the world (Figure 3.97). Adults are 6–9 mm long, black with a blue head, and have four large, yellow spots and a reddish margin on their abdomen. Both adults and larvae feed on the young shoots. The 12-spotted asparagus beetle (*C. duodecimpunctata*) is similar in size, but bright orange with black spots; adults produce minor damage to the foliage while their larvae feed mainly on the red asparagus berries. Eggs of both beetles are shiny dark brown to black in color and oblong shaped. They hatch to a larva (gray/white for *C. asparagi* and orange for *C. duodecimpunctata*) that

Figure 3.94   A wireworm or click beetle larva (*Agriotes* spp.).

Figure 3.95   An adult turnip flea beetle (*Phyllotretes nemorum*).

Figure 3.96   Adult flea beetles (*Phyllotreta* spp.) and their feeding damage on young brassica.

feeds on asparagus foliage (Figure 3.98) before dropping to the ground, where it continues feeding on stems before pupating. There may be three or more generations of these beetles each year.

Figure 3.97   An adult asparagus beetle (*Crioceris asparagi*) on an asparagus fern.

Colorado potato beetle is one of the most infamous Chrysomelidae (Figures 3.99 and 3.100) and may be found across North America and most of Europe but not Britain or Ireland. Its principal food plant is the potato, but severe damage can occur on aubergine, pepper, and tomato. Both larvae and adults feed on leaves of solanaceous plants, which may result in plant death. Colorado potato beetle is a voracious feeder and, in high numbers, can cause considerable leaf loss (Figure 3.101), leading to greatly reduced yields. Colorado beetle is a notorious pest due to its ability to become resistant to pesticides applied for its control.

Predatory shield bugs of *Podisus* spp. are being used as biocontrol organisms in several countries with good success. Similarly, specific

Figure 3.99   An adult Colorado beetle (*Leptinotarsa decemlineata*) on potato.

Figure 3.98   An asparagus beetle larva (*Crioceris asparagi*) on an asparagus fern.

Figure 3.100   A late larval stage of a Colorado potato beetle (*Leptinotarsa decemlineata*) on potato.

Figure 3.101 Potato haulm destroyed by adult Colorado potato beetles (*Leptinotarsa decemlineata*).

Figure 3.102 Cabbage or winter stem weevil larvae (*Ceutorhynchus quadridens*) in a damaged oilseed rape or canola stem.

strains of *Bacillus thuringiensis* and *Beauveria bassiana* are widely used against this pest.

WEEVILS

This is an important family of beetles and deserves special mention due to the widespread nature of damage done to many agricultural (Figure 3.102), nursery, soft fruit, and garden plants (Figure 3.103 and 3.104). Many species are native to Europe and are an introduced pest to North America, where damage ranges from troublesome to severe. *Otiorhynchus sulcatus* is commonly known as the black vine weevil, cyclamen borer, or strawberry root weevil and is found on many horticultural crops throughout the temperate regions of the world. Other species, such as *O. singularis* and *O. rugosostriatus* (clay colored and lesser strawberry weevil, respectively), have similar life cycles but are smaller than *O. sulcatus* at all stages.

Economic damage can be produced by a single weevil larva feeding on a cyclamen corm, whereas several larvae may kill or weaken quite mature plants. Adults are nocturnal feeders on leaves and produce a characteristic edge notching that, in itself, may cause economic damage to ornamental plants (Figure 3.105). In outdoor or unheated crops, there is usually only one generation each year, but in heated conservatories, botanic gardens,

Figure 3.103 A cabbage or winter stem weevil adult (*Ceutorhynchus quadridens*) on an oilseed rape or canola leaf.

Figure 3.104 An adult black vine weevil (*Otiorhynchus sulcatus*).

Figure 3.105   Adult vine weevil (*Otiorhynchus sulcatus*) feeding damage to *Viburnum davidii* foliage.

Figure 3.106   Two larvae and a pupa of black vine weevil (*Otiorhynchus sulcatus*).

and interior atriums there may be a number of overlapping generations. All recorded adult vine weevils are female, so reproduction is entirely parthenogenetic. This is likely a result of a heritable, universal *Wolbachia* bacterial infection that affects the ovaries, thus leading to the elimination of males within the species. These bacteria are found in the reproductive tissues of many arthropods and are transmitted through the cytoplasm of their eggs. Infected organisms may all become female (feminization) or sterile or show reproductive incompatibility.

Adults are wingless and 7–10 mm in length. Migration is limited when compared with many other insects, although plant movement by man accounts for most rapid spread. Eggs, initially soft white and glistening, are laid on the compost surface and a few days later they melanize to become rigid and reddish-brown in color. The newly hatched or neonate larva is white and legless with a brown head. Older larvae are normally C-shaped and may acquire some color from their host

plant tissue (staining). Mature larvae may begin to burrow into corms and fleshy stems of various plant species. While in the stem, they are well protected from most forms of control including pesticides. Larvae overwinter in compost or soil within cells produced by wriggling around to create a small pocket where they remain dormant until warmer conditions return and feeding resumes. After the final larval instar is reached, a similar pupal cell is formed in which the weevil pupates (Figure 3.106), emerging some weeks later as an adult.

Natural enemies of vine weevil include insectivorous mammals (hedgehogs and moles), some bird species, frogs, toads, lizards, and predatory insects such as carabid beetles. Entomopathogenic fungi and nematodes are also natural enemies. Commercial biocontrol agents include the fungi *Beauveria bassiana* and *Metarhizium anisopliae* and nematodes *Heterorhabditis bacteriophora*, *H. megidis*, *Steinernema carpocapsae*, and *S. kraussei*.

Arthropod biological control agents

INTRODUCTION

In this chapter many of the terrestrial parasitoid and predatory arthropods are described. They occur in a wide range of situations and in many different crops. The groups are considered according to their taxonomic order (Table 4.1). For each arthropod, a description is given of its characteristics with details of each stage in its life cycle. The habitat of the arthropod is described, together with the pest organisms with which it associates and, finally, the influence of growing practices on the beneficial.

Several parasitoid wasps are specific to their host—even to the stage of the host they attack. In this respect they may have a narrow window of opportunity in which to reproduce. Solitary species deposit one egg per host (e.g., *Aphidius* spp. and *Encarsia* spp.), while gregarious species can produce several parasitoids from each host (e.g., *Cotesia glomerata*).

Predatory wasps tend to be less specialized in their food requirements; many are polyphagous, feeding on prey in different families and at different life stages (e.g., the larvae of *Chrysoperla carnea* that can feed on several soft-bodied prey). Many predators do not need to synchronize with their prey; provided sufficient prey is present in the immediate area, they will continue to develop and reproduce. This is a particularly useful trait for commercially produced predators that may be utilized as living pesticides. Many predators are carnivorous as both adult and nymph (e.g., *Macrolophus pygmaeus*). Similarly, many of the ground beetles are predatory as adults and larvae, whereas the Neuroptera and Syrphidae have only predatory larvae.

One common although arbitrary factor is that this group of organisms is visible to the naked eye at the adult and usually the later larval or nymphal stages. Although adult parasitoids may be visible, many develop within a host that must be dissected for the parasitoid to be observed at the larval stage. The majority of beneficial organisms, whether naturally occurring or commercially

TABLE 4.1 Arthropod biological control agents

Subphylum (SP), class (C), subclass (SC), division (D)	Order and suborder (SO)	Family	Genus and species	Common name	Page
(SP) Chelicerata (C) Arachnida	Araneae	Araneidae	*Araneus diadematus*	Orb web spiders	70
		Linyphiidae	*Erigone atra*	Money spiders	72
		Lycosidae	*Pisaura mirabilis*	Wolf spiders	73
		Salticidae	*Salticus scenicus*	Jumping spiders	74
		Thomisidae	*Misumena vatia*	Crab spiders	75
	Opiliones	Phalangiidae	*Phalangium opilio*	Harvestmen	75
(SC) Acari	Prostigmata	Trombidiidae	*Allothrombidium* spp.	Velvet mites	77
	Mesostigmata	Phytoseiidae	*Amblyseius andersoni*	Predatory mites	78
			Amblyseius californicus	Predatory mites	79
			Amblyseius cucumeris	Predatory mites	80
			Amblyseius degenerans	Predatory mites	81
			Amblyseius montdorensis	Predatory mites	82
			Amblyseius swirskii	Predatory mites	82
			Phytoseiulus persimilis	Spider mite predator	84
			Typhlodromus pyri	Spider mite predator	85
		Laelapidae	*Hypoaspis aculeifer*		86
			Hypoaspis miles	Sciarid larva predator	86
		Macrochelidae	*Macrocheles robustulus*	Sciarid larva predator	87
(SP) Myriapoda					
(C) Chilopoda	Geophilomorpha	Geophilidae	*Geophilus flavus*	Centipede	88
	Lithobiomorpha	Lithobiidae	*Lithobius forficatus*	Centipede	88
			Stigmatogaster subterraneus	Centipede	88
(SP) Hexapoda					
(C) Insecta					
(SC) Pterygota					
(D) Exopterygota	Odonata				
	(SO) Anisoptera	Aeshinidae	*Aeshna cyanea*	Dragonfly	89
	(SO) Zygoptera	Coenagrionidae	*Coenagrion puella*	Damselfly	90
			Pyrrhosoma nymphula	Damselfly	91
	Dermaptera	Forficulidae	*Forficula auricularia*	Earwigs	93

Continued

TABLE 4.1 *Continued*

Subphylum (SP), class (C), subclass (SC), division (D)	Order and suborder (SO)	Family	Genus and species	Common name	Page
	Hemiptera			Predatory bugs	95
	(SO) Heteroptera	Gerridae	*Gerris lacustris*	Pond skaters	96
		Anthocoridae	*Anthocoris nemoralis*		97
			Anthocoris nemorum		98
			Orius laevigatus	Thrips predator	99
			Orius majuscules	Thrips predator	100
		Miridae	*Atractotomus mali*	Predatory bugs	101
			Blepharidopterus angulatus	Predatory bugs	101
			Deraecoris ruber	Predatory bugs	102
			Heterotoma planicornis	Predatory bugs	103
			Macrolophus pygmaeus	Predatory bugs	104
			Pilophorus perplexus	Predatory bugs	105
		Pentatomidae	*Podisus maculiventris*	Caterpillar predator bug	106
	(SO) Sternorrhyncha	Psyllidae	*Aphlara itadori*		107
	Thysanoptera	Thripidae	*Franklinothrips vespiformis*	Predatory thrips	109
(D) Endopterygota					
	Neuroptera	Chrysopidae	*Chrysopa perla*	Green lacewing	111
			Chrysoperla carnea	Green lacewing	111
	Coleoptera			Beetles	113
		Coccinelidae		Ladybird beetles	113
			Chilocorus bipustulatus		116
			Coccinella septempunctata	Seven-spot ladybird	117
			Cryptolaemus montrouzieri	Mealybug predator	119
			Delphastus catalinae		120
			Harmonia axyridris	Harlequin ladybird	121
			Harmonia congloblata		123
			Hippodamia convergens	Convergent ladybird	124
			Propylea 14-puncata	Fourteen-spot ladybird	125
					Continued

TABLE 4.1 *Continued*

Subphylum (SP), class (C), subclass (SC), division (D)	Order and suborder (SO)	Family	Genus and species	Common name	Page
			Rodolia cardinalis	Vedalia beetle	126
			Scymnus subvillosus		127
			Stethorus punctillum		127
		Carabidae		Ground beetles	128
			Agonum dorsale		130
			Bembidion lampros		131
			Carabus violaceus	Violet ground beetle	131
			Demetrias atricapillus		132
			Harpalus rufipes		133
			Loricera pilicornis		134
			Nebria brevicollis		134
			Notiophilus biguttatus		135
			Poecilus cupreus		136
			Pterostichus melanarius		136
			Trechus quadristriatus		137
		Cicindelidae	*Cicindela campestris*	Field tiger beetle	138
		Staphylinidae	*Aleocharina* spp.	Rove beetles	139
			Atheta coriaria	Rove beetles	140
			Philonthus cognatus	Rove beetles	141
			Tachyporus spp.	Rove beetles	142
			Xantholinus spp.	Rove beetles	143
	Diptera	Empididae	*Empis sterocorea*	Dance flies	144
			Empis tessellate	Dance flies	144
		Muscidae	*Coenosia attenuata*	Hunter flies	145
		Tachinidae	*Tachinid* spp.	Parasitic flies	146
		Cecidomyiidae	*Aphidoletes aphidimyza*	Aphid predatory midge	147
			Feltiella acarisuga	Spider mite	149
		Syrphidae	*Episyrphus balteatus*	Aphid hoverfly	152
			Scaeva pyrastri	Aphid hoverfly	154
			Syrphus ribesii	Aphid hoverfly	155
	Hymenoptera			Parasitoid wasps	156
		Mymaridae	*Anagrus atomus*	Leaf hopper egg parasitoid	159

Continued

TABLE 4.1 *Continued*

Subphylum (SP), class (C), subclass (SC), division (D)	Order and suborder (SO)	Family	Genus and species	Common name	Page
		Aphelinidae	Aphelinus abdominalis	Aphid parasitoid	159
			Encarsia formosa	Whitefly parasitoid	161
			Encarsia tricolor	Whitefly parasitoid	163
			Eretmocerus eremicus	Whitefly parasitoid	164
		Braconidae	Aphidius colemani	Aphid parasitoid	166
			Aphidius ervi	Aphid parasitoid	167
			Cotesia glomerata	Caterpillar parasitoid	168
			Dacnusa sibirica	Leaf miner parasitoid	169
			Diaeretiella rapae	Aphid parasitoid	170
			Praon myziphagum	Aphid parasitoid	171
			Praon volucre	Aphid parasitoid	171
		Ichneumonidae	Diadegma insulare	Caterpillar parasitoid	172
		Eulophidae	Diglyphus isaea	Leaf miner parasitoid	173
			Tetrastichus asparagi	Asparagus beetle wasp	175
		Encyrtidae	Encyrtus infelix	Soft scale insects parasitoid	176
			Leptomastix dactylopii	Mealybug parasitoid	177
			Leptomastix epona	Mealybug parasitoid	177
			Metaphycus helvolus	Soft scale insects wasp	178
		Tricho-grammatidae	Trichogramma spp.	Moth egg parasitoid	179
		Crabronidae	Cerceris arenaria	Predatory wasp	180
		Vespidae	Vespula vulgaris	Predatory wasp	181

produced, are susceptible to pesticides, particularly long-persistence, synthetic insecticides.

CLASS: ARACHNIDA (SPIDERS, HARVESTMEN, AND PREDATORY MITES)

Spiders, or arachnids, are present in one form or another as polyphagous predators in virtually all crops and gardens. They can be extremely numerous but often go unnoticed because of their size and relatively secretive habits. Different spider families have different techniques for catching their prey, with some spinning silken webs (Figure 4.1) and others jumping or running. Most will bite if handled roughly, but few are venomous to mammals.

Like other arthropods, the arachnids have an external skeleton and a body made up of a series of segments, each characterized by an internal nerve ganglion and a pair of jointed appendages. The body is divided into

Figure 4.1 European garden spider (*Araneus diadematus*) on a dew-covered web in the early morning.

Figure 4.3 A female *Trochosa* spp. wolf spider producing silk egg sac.

Figure 4.2 Silk balloon threads produced by Linyphiid money spiders.

the contents to be sucked out. Spiders have six (or, more typically, eight) eyes. The relative position of the eyes is widely used in the classification of spiders. However, spiders are difficult to identify readily in the field and it is often necessary to examine them with a hand lens or a binocular microscope to classify them more precisely. Male spiders are typically identified by the shape of their palps (Figure 4.4), whereas females are normally identified by their epigynes (the female genital area on the underside of the abdomen). Color and size are not always reliable features when identifying spiders.

Family: Araneidae (Orb Web Spiders)

Orb web spiders have relatively poor eyesight and are all spinners of orb-shaped webs. They are usually seen sitting on or to one side of the web and detect the presence of prey in the web by vibration. When prey is caught in the web, it is wrapped in silk before being bitten. A new web may be constructed almost daily, with the remnants of damaged webs consumed for recycling. The common or European garden spider, *Araneus diadematus* (Figure 4.5), is found throughout Europe and Asia; it is also reported in several states of America. Females can be up to 15 mm in length, while males are smaller at 8–9 mm; colors range from yellow, brown, to dark red—often in a camouflage pattern. One of the most common orb web spiders in

two parts. The anterior cephalothorax comprises seven segments with the first having only the brain-like ganglia and no appendages; the remaining six segments have a pair of large piercing chelicerae and then a pair of sensory pedipalps, followed by four segments each bearing a pair of walking legs. The rear end of the body, called the opisthosoma, or abdomen, comprises another series of segments that generally have no appendages but have openings for respiration or reproductive ducts and silk-spinning organs (Figures 4.2 and 4.3).

Spiders are characterized by having a narrow waist so that the rear end has the mobility to wrap the prey in silk and to spin elaborate webs. Most arachnids feed only on fluids; in particular, spiders inject a paralyzing venom and a digestive juice into the prey that breaks down the body structure, enabling

Figure 4.4   Head view of *Araneus diadematus* showing eyes and palps.

Figure 4.5   Close-up of European garden spider (*Araneus diadematus*) on a dew-covered web in the early morning.

fruit orchards is *Araniella cucurbitin*. It has a bright green abdomen with paired black spots. The females are 4–6 mm in length and the males are 3.5–4 mm in length.

- *Life cycle:* The life cycles of the Orb web spiders vary with species.
- *Crop/pest associations:* Orb web spiders feed on insects that fly into their webs (Figure 4.6). In fruit orchards these may include pest species such as aphids or psyllids. Because they are polyphagous and are not restricted to a particular prey type, they can survive in crops with no pest species present.
- *Influence of growing practices:* Orb webs are common in apple and pear orchards and have been shown to be more abundant where flowering strips are also present. The flowering plants attract a wide range of flying insect species, many of which are trapped in their webs.

Figure 4.6   Close-up of *A. diadematus* repairing orb-shaped web.

Orchards planted with flowering strips have been shown to have fewer aphids where higher numbers of Orb web spiders were observed. Because web-spinning spiders may reuse parts of their webs over several months, they may acquire higher doses of pesticides, particularly long-persistence synthetic neonicotinoid and pyrethroid insecticides—potentially making them more sensitive to pesticide sprays than other spiders.

Family: Linyphiidae (Money Spiders)

This is one of the largest families of spiders and they are the most numerous in gardens and agricultural crops. Money spiders are difficult to identify since they are classified by the absence of the characteristics of other spider families. They are plainly marked and considered to bring good fortune—hence the name money spiders. The most common genera in agricultural crops include *Erigone* (Figures 4.7 and 4.8), *Oedothorax* (Figure 4.9), and *Lepthyphantes* (Figures 4.10 and 4.11). They range in size from 1 to 5.5 mm.

- *Life cycle:* Eggs are laid in egg sacs on the underside of leaves or under stones. Webs are constructed but with no refuge, and the spiders are often seen on the underside of their webs. In late summer and early autumn,

Figure 4.7   Adult money spider (*Erigone* sp.).

Figure 4.8   Adult female *Erigone atra,* a money spider.

Figure 4.9   Adult *Oedothorax retusus,* a money spider.

Figure 4.10   *Lepthyphantes tenuis,* a linyphiid money spider on cereal.

money spiders can be found dispersing on threads of silk and being blown on currents of warm air ("ballooning").

- *Crop/pest associations:* They are polyphagous predators in all crops and gardens and consume aphids and other invertebrate prey.
- *Influence of growing practices:* They are very numerous in most crops and gardens and spin webs at different heights within the canopy. Aphids, which regularly fall from the plant to the ground, can be caught in these webs, as can smaller flying insects.

Figure 4.11 Adult *Lepthyphantes tenuis* on ear of wheat.

Figure 4.12 Wolf spider (Lycosidae) webbing on heather covered with early morning dew.

Figure 4.13 A wolf spider (*Pisaura mirabilis*) with web on wheat.

Family: Lycosidae (Wolf Spiders)

These are the most common large spiders found on the ground in most agricultural crops (Figure 4.12). They are identified by the position of their eyes with the front of the carapace having a row of four small eyes, evenly spaced. Above them is a pair of larger forward-facing eyes. Further back on the carapace is a second pair of large eyes. A line through the outer large eyes will meet the midline of the spider in front of the carapace. These spiders are generally brown in color with markings on the carapace and abdomen and with a dense covering of hairs (Figure 4.13). Some are active at night, but many are also seen running on the soil surface on warm, sunny days. Species are identified by examination of the palps of males and the epigynes of the females.

- *Life cycle:* From early summer onward, female spiders begin to carry a conspicuous pale colored egg sac attached to their abdomen (Figure 4.14). After 2–3 weeks the eggs hatch and the young spiderlings are carried by the female for up to 1 week. Females may make up to three egg sacs in a season. Males generally die in midsummer.
- *Crop/pest associations:* They are hunting spiders and will feed on a variety of small insects or mites in most crops. Aphids are

Figure 4.14   Female *P. mirabilis* rolling egg sac to protective position.

Figure 4.15   A *Pardosa* spp. wolf spider with egg sac (at rear).

Figure 4.16   Male zebra jumping spider (*Salticus scenicus*).

likely to enhance wolf spider numbers. They are sensitive to insecticides, particularly neonicotinoids, synthetic pyrethroids, and organophosphates.

Family: Salticidae (Jumping Spiders)

These are hunting spiders with very good eyesight and have a characteristic square-fronted carapace with four large, forward-facing eyes on the front, a smaller pair slightly further back, and a larger pair further back still. If an object enters their field of vision, they will jump round to focus the large front eyes on it. Jumping spiders are often covered in hairs, which may be brightly iridescent or dark (Figure 4.16). They do not spin webs and several species are thought to mimic ants. These cosmopolitan spiders are called jumping spiders because they often jump to capture their prey.

- *Life cycle:* The life cycles of jumping spiders are diverse, even within a genus.
- *Crop/pest associations:* These spiders are polyphagous predators, consuming a wide range of insect prey in almost every crop system. They prefer ground cover and are often seen running through grass and low vegetation. Some species are also active on tree trunks.
- *Influence of growing practices:* Jumping spiders are unlikely to be active in areas with bare soil—for example, on herbicide-treated strips beneath

eaten when on the ground; the spiders do not actively climb the plants in search of prey. In cereal crops, spiders in the genus *Pardosa* (Figure 4.15) are particularly common and are voracious polyphagous predators.

- *Influence of growing practices:* Wolf spiders like shelter and high humidity and are less abundant in open, dry habitats. Provision of beetle banks is orchard trees. They are more abundant in unsprayed crops and are particularly susceptible to pyrethroid insecticides.

Family: Thomisidae (Crab Spiders)

Crab spiders have the front two pairs of legs longer than the back two pairs and usually adopt a characteristic crab-like stance. These spiders can move sideways like a crab, as well as forward and backward. There can be considerable color variation within species and they are often colored to match their surroundings (Figures 4.17 and 4.18). They do not spin webs but rather catch their prey by waiting camouflaged in flowers or on leaves with their front legs held open. When an insect ventures near enough the spider pounces and seizes its prey. *Xysticus cristatus* is the most common of these spiders and is found on trees and bushes. It has a pattern of stacked triangles on its abdomen. Adult spiders typically have body lengths of between 4 and 8 mm.

- *Life cycle:* Crab spiders have varied life cycles, even within a genus.
- *Crop/pest associations:* They are common as predators in fruit crops, particularly in warmer climates, and are polyphagous, feeding on a variety of arthropod species including insect pests. They are particularly noticeable when there are large numbers of flying insects in a crop—for example, at the time of flowering in fruit trees.
- *Influence of growing practices:* Like all spiders these are extremely sensitive to pyrethroid insecticides but are probably not harmed by most fungicide treatments.

Figure 4.17   Goldenrod crab spider (*Misumena vatia*) waiting for prey.

Figure 4.18   A pale colored *Misumena vatia* camouflaged against petals.

Family: Phalangiidae (Harvestmen)

Harvestmen such as *Phalangium opilio* have four pairs of legs that are unusually long in relation to their body (4.0 to 9.0 mm in length), which, seen from above, is oval in shape (Figure 4.19). It is quite common for this group of spiders to lose one or more legs (Figures 4.20 and 4.21) during their life, causing little or no problem with their hunting and survival. This group has the same body structure described for spiders but lacks the narrow waist and the ability to produce silk. Harvestmen have two sideways-facing eyes, which are generally mounted on a central raised structure called an ocularium (Figure 4.22). Females are larger than males.

- *Life cycle:* Eggs are laid in batches in soil or other damp medium during the late summer and autumn and hatch in

Figure 4.19   **Adult female *Phalangium opilio* with seven legs; note harvestmen commonly lose one or more legs with little negative impact.**

Figure 4.20   **Adult *Leiobunum rotundum* extending one of its legs as a feeler.**

Figure 4.21   **Adult female *Leiobunum rotundum* with seven legs.**

Figure 4.22   **Adult female *Oligolophus tridens* on apple leaf.**

Figure 4.23   **Adult male *Dicranopalpus ramosus* lying in wait for prey.**

the spring. Juvenile harvestmen may pass through eight instars during spring and summer and reach sexual maturity in late summer. They are most abundant in midsummer, but adults and juveniles of *P. opilio* can also overwinter.

- *Crop/pest associations:* Harvestmen are found throughout the Palearctic region, North America, and New Zealand. They are predators and occur in most croplands and gardens where there is high humidity. *P. opilio* occurs in disturbed habitats among grass and surface litter. Harvestmen will attack any available soft-bodied insects including aphids (Figure 4.23).
- *Influence of growing practices:* Harvestmen cannot tolerate dry conditions and prefer ground close to water or shelter. Grassy banks at field margins will encourage them, as will dense undergrowth and hedgerows.

SUBCLASS: ACARI (PREDATORY MITES)

These are small, fast moving mites that can be specific predators, as with *Phytoseiulus persimilis* (Figure 4.24), or more generalist in their diet, as are many of the *Amblyseius* species. In this book we have retained the long established generic names but have also shown the newly proposed names.

All deposit eggs close to the intended prey that hatch as six-legged nymphs, pass through two molts, and develop as eight-legged adults. Location of prey is usually by kairomones released by the prey feces, plant damage, or, in the case of spider mites, by their webbing that produces an attractant and arrestment stimulus in the predator that retains them in close vicinity of their host prey. Most predatory mites are capable of surviving on relatively low numbers of prey and can increase rapidly to provide adequate levels of control before any major outbreak occurs. Predatory mites are found throughout the world and several are in commercial production for mass release to many crops, particularly protected crops. However, their use on outdoor crops is increasing, specially on edible crops where post-harvest intervals of many pesticides restrict or even prevent chemical intervention. Many of the

Figure 4.25   Long lengths of paper sachets containing predatory *Amblyseius* mites and a factitious prey mite as food are easily distributed among many different crops—in this instance, a commercial alstromeria crop in Cornwall, UK.

Amblyseied mites can be mass produced on a bran-based diet that may be packaged along with a factitious host mite in paper sachets (Figure 4.25) for ease of distribution and improved establishment on a crop.

Family: Trombidiidae (Red Velvet Mites)
ALLOTHROMBIDIUM SPP.

These are commonly known as red velvet mites due to high levels of carotene, which gives a bright red color, and the velvet coat of fine hairs covering their bodies (Figure 4.26). *Allothrombidium* spp. are the largest of this mite family. Adult females are 4 mm in length and become more noticeable under hot, dry conditions, when they run freely over the ground, including stone and concrete surfaces. Velvet mites are common throughout temperate regions; tropical species of giant red velvet mites can reach 10–12 mm × 6–8 mm across and are specialized predators of termites.

- *Life cycle: Allothrombidium* spp. have one generation each year and spend most of their life underground, only venturing out during warm, dry periods. Small, orange-colored eggs, 0.2 mm in diameter, are deposited in crevices below the ground and hatch to a six-legged larva that attaches itself to a host

Figure 4.24   Predatory mite (*Phytoseiulus persimilis*) (orange color) attacking prey mite (*Tetranychus urticae*).

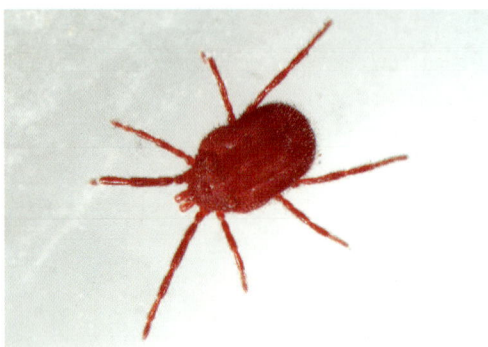

Figure 4.26   Adult red velvet mite (*Eutrombidium rostratus*).

Figure 4.27   Velvet mite (*Allothrombidium fuliginosum*) parasitic nymph attacking (*Macrosiphum euphorbiae*).

prey. After its first molt, the larva gains another pair of legs to form an eight-legged nymph.

- *Crop/pest associations:* The larvae are ectoparasitoids and feed on many hosts, including insects and other mites. Adults and nymphs are predatory and are frequently found among aphid colonies (Figure 4.27), where they may kill several individuals per day.
- *Influence of growing practices:* Frequently cultivated ground can upset their breeding sites. Red velvet mites are thought to have few natural enemies.

Family: Phytoseiidae (Predatory Mites)

AMBLYSEIUS ANDERSONI

Amblyseius andersoni is a predatory mite that feeds on small prey organisms such as gall mites, rust mites, spider mites, and

thrips. In the absence of arthropod prey, they can survive on fungal spores, pollen, and plant exudates. They are widespread in Europe and North America, where they are found on fruit crops such as apple, grape, peach, and raspberry. Several rose cultivars have extrafloral nectaries at the base of leaves that provide additional food for predatory mites, allowing good establishment before pest numbers increase. Extrafloral nectaries are also found on cherry, cotton, cowpea, peach, zucchini, and various ornamental trees.

- *Life cycle:* Adult female mites attach single eggs onto leaf hairs, which hatch after 2–3 days to produce six-legged larvae that molt to produce a protonymph, followed soon after by a deutonymph. All mobile stages are predatory and will feed on eggs, juveniles, and adults of their prey. The mites enter diapause in early autumn to overwinter in protected cracks and crevices, becoming active again in late winter when ambient temperature and plant growth increase. It is possible to mass produce *A. andersoni* in a bran-based medium using stored product mites (Figure 4.28) as food and to distribute them in paper sachets, which are carefully designed to maintain the colony in good condition for

Figure 4.28   Predatory mite *Amblyseius andersoni* (oval shape) attacking food storage mite (*Acarus siro*; brown front legs) as used in mass-production systems.

several weeks. These are commercially available in many countries.

- *Crop/pest association:* As polyphagous feeders they can establish on many plants much better than specialized predators, such as *Phytoseiulus persimilis*, that feed on just one food source (spider mites). *A. andersoni* mites are introduced for biological control of spider mites in many crops and can be supplemented with *P. persimilis* to control larger infestations. They are also commercially used against several other mite pests. The use of paper sachets (Figure 4.29) containing predator and prey mites gives a higher and more prolonged release rate than distribution in loose bran or a vermiculite carrier. They are mostly used as a preventive measure on susceptible crops, but can be difficult to find on plants after release without using leaf-washing techniques.

- *Influence of growing practices:* Wild populations of *A. andersoni* are reported to have a degree of pesticide tolerance and can be one of the first predatory mites to invade fruit crops after pesticide applications are reduced. For example, strains tolerant to pyrethroid insecticides have been reported from Italy. The sachet used to release predatory mites will protect the colony from insecticidal treatments, so if a short-persistence spray is used and mites on the foliage are killed, fresh mites will emerge from the sachets to replace them. This works well with short-persistence, physically acting contact pesticides and those with known selectivity to predatory mites such as spinosad. Wild plants such as *Selene dioica* (red campion), which has hairy leaves and produces a good supply of nectar and pollen, act as reservoirs and hibernation sites for several species of predatory mites. These plants are frequently available from garden centers and nurseries supplying flowering plants.

AMBLYSEIUS CALIFORNICUS (SYN. NEOSEIULUS CALIFORNICUS)

Amblyseius californicus is a predatory mite specialized in feeding on spider mites (Figure 4.30), including *Panonychus ulmi* (fruit tree spider mite), *Tetranychus urticae* (glasshouse spider mite), and *T. cinabarinus*

Figure 4.29 Controlled release system (CRS) sachet placed in commercial crop to introduce amblyseid predatory mites.

Figure 4.30 Predatory mite (*Amblyseius californicus*) attacking young two-spotted spider mite (*Tetranychus urticae*).

(carmine spider mite). Adults are 1.25 mm in length and can produce over 60 eggs per female. They can also feed on other arthropod prey and pollen when the favored food is scarce. This allows better survival than many obligate predators.

This species is commercially available in many countries, but a license for release is required in some.

- *Life cycle:* Adult females attach their eggs singly to leaf hairs along veins on the undersides of leaves close to spider mite colonies. These hatch to a six-legged larva, which develops through to protonymph and deutonymphal stages before reaching adulthood. This takes almost 10 days at 21°C (70°F) but only 5 days at 30°C (86°F), which is marginally quicker than the cycle of its principal prey, *T. urticae*.
- *Crop/pest associations:* Studies on Californian strawberries indicate that *A. californicus* is a less effective predator than *Phytoseiulus persimilis* because fewer eggs are laid. It has a lower rate of population growth, searching for prey is less thorough, and fewer prey are consumed. However, it is reported to persist longer in various glasshouses and field crops and then to control spider mites in situations where *P. persimilis* is not so effective. These include hot, dry environments and on ornamental or foliage plants where spider mite densities may be low.
- *Influence of growing practices:* *A. californicus* is reported as being highly tolerant to several of the more commonly used pesticides from the organophosphate, carbamate, and synthetic pyrethroid groups. In any environment where predators are being used, it is always advisable to use selective or compatible pesticides, such as those that are physically active, the majority of which cease activity once the spray has

dried. Due to the slower rate of activity of the biocontrol agent, a selective or short persistence acaricide may be necessary to reduce plant damage before the predator gives adequate levels of control.

AMBLYSEIUS CUCUMERIS (SYN. *NEOSEIULUS CUCUMERIS*)

Amblyseius cucumeris is a predatory mite and one of a group of thrips feeders. It is 1.25 mm in length and pale in color. Nymphs and adults have eight legs and use the front pair as feelers. They are polyphagous predators and naturally occurring in many crops throughout the temperate regions. Their commercialization has increased their geographical distribution.

They are commercially available in many countries.

- *Life cycle:* The oval eggs (about 0.14 mm diameter) are deposited on pubescent hairs of the midvein axil and similarly on the lateral veins on the undersides of leaves. They are translucent pink/white in color. On hatching the larva have six legs and do not feed for several hours. The two nymphal stages both have eight legs and are very mobile and active feeders; adults behave similarly. Commercially, they are reared on fine bran flakes (Figure 4.31) infested with

Figure 4.31 Predatory mite (*Amblyseius cucumeris*) with flakes of bran as used to mass produce these mites.

a flour mite such as *Acarus siro* or another stored-product mite such as *Tyroglyphus* spp.

- *Crop/pest associations: A. cucumeris* are introduced to the crop with the rearing medium, usually diluted with a carrier material such as vermiculite granules, placed in shaker bottles or tubes. Pure bran flakes infested with the prey mite and predator are used in sachets to provide a continuous release of predators over a 6- to 8-week period. These sachets are placed on growing plants and over a period of weeks effectively colonize many leaves in the surrounding area. If the product is examined closely, the slow moving, hairy, white flour mites can be seen and may be mistaken for the predator. The predator attacks mainly the first instar thrips larva, so it is necessary to maintain an active population of mites on plants to gain maximum levels of pest control. It is quite usual to introduce *A. cucumeris* as a preventive treatment. However, on plants that flower continuously and produce pollen, such as sweet pepper, the mite will readily establish from a single introduction.

- *Influence of growing practices: A. cucumeris* is susceptible to most insecticides and several fungicides used in crop protection. However, mites released from sachets have more protection and are generally effective when used with short-persistence, physically acting pesticides. Low humidities early in the growing season can disrupt the development of this predatory mite, requiring repeat introductions. This is one of the cheapest predators to buy in large numbers. They are used as a "living pesticide" on several ornamental pot plant crops to give continuous protection when introduced weekly at rates of between 100 and 300 mites/m².

Figure 4.32   Predatory mites (*Amblyseius degenerans*) on leaf surface.

*AMBLYSEIUS DEGENERANS (*SYN. *IPHISEIUS DEGENERANS)*

This is a shiny black, oval-shaped mite, 1 mm in length, that rapidly runs over leaves (Figure 4.32). *Amblyseius degenerans* originates in the Mediterranean region and North Africa. It feeds on a wide range of food including thrips larvae, spider mites, and other small arthropod prey, which it sucks dry. The mites are able to survive and reproduce on a diet of pollen, but when food is limited, they can be strongly cannibalistic, with females eating males and all immature stages.

This mite is commercially available in many countries, but a license for release is required in some.

- *Life cycle:* Adult females deposit eggs on leaf hairs, usually at the junction of veins and at curled leaf edges. These are initially transparent but turn a brown color as they mature. Unusually, the mites can lay their eggs onto those laid earlier, even when eggs were laid by other females, giving rise to quite large egg clusters at favored sites. Eggs hatch after a couple of days at 25°C (77°F) and can develop to adulthood within a week at this temperature. Male mites attach themselves to female protonymphs and mating occurs almost immediately as the young female emerges.

- *Crop/pest associations:* A. degenerans establishes well where there is a continuous production of pollen and the leaf surface is smooth. This includes such plants as citrus species and sweet pepper. To prevent the establishment of thrips, *A. degenerans* can be introduced by inoculation to young plants once they have begun to flower and produce pollen. Trials in Holland have used banker plants of castor oil (*Ricinus communis*) for the production of this predator. These are placed close to rows of sweet pepper plants. The castor oil leaves are then periodically removed for placement on the crop in areas of pest activity.

- *Influence of growing practices:* A. degenerans is sensitive to many fungicides and insecticides in common horticultural use. The vaporization of sulfur to control powdery mildew causes a reduction in their numbers. On pepper, it will attack larger thrips (unlike *A. cucumeris,* which is restricted to first instar thrips larvae) and tolerate lower humidities. This mite has not been shown to control thrips populations on any crop other than sweet pepper.

*AMBLYSEIUS MONTDORENSIS (*SYN. *TYPHLODROMIPS MONTDORENSIS)*
Approximately 1 mm in length, the predatory mite *Amblyseius montdorensis* (Figure 4.33) was first described in 1979 in subtropical

Figure 4.33   Adult *Amblyseius montdorensis* predatory mites.

areas such as Fiji, New Caledonia, New Hebrides, Tahiti, and Queensland, Australia. This generalist feeder may be found attacking small arthropods such as rust and gall mites (Eriophyidae), tarsonemid mites, spider mites, and thrips.

It is commercially available in many countries, but a license for release is required in some.

- *Life cycle:* Female mites lay eggs on the leaf surface. These hatch after a day or so to produce six-legged larvae that pass through two nymphal stages to reach adulthood in under a week at 25°C (77°F).

- *Crop/pest associations:* Adults and older nymphs disperse well through plant canopies and even between plants in search of pollen or prey. This voracious predator has a high reproductive rate and can readily establish on several leaf and plant types (several species of predatory mites do not establish well on hairy leaves). Current rates of introduction are two applications of between 5 and 20 mites/m^2 compared to *A. cucumeris,* which is normally introduced at 100–300 mites/m^2 weekly, depending on the crop.

- *Influence of growing practices:* This predatory mite is still under evaluation in much of Europe and North America. However, studies have indicated that development of *A. montdorensis* is reduced at temperatures below 15°C (63°F) and humidities below 60%. Even in crops with a high daytime temperature, it may not establish during winter and spring.

*AMBLYSEIUS SWIRSKII (*SYN. *TYPHLODROMIPS SWIRSKII)*
Amblyseius swirskii (Figure 4.34) is a small predatory mite that feeds on many types of small arthropod prey and pollen. It originates from the Nile Delta of Egypt and adjacent areas of Israel. It requires high

take less than 7 days. In optimum conditions, adults can produce more than two eggs per day, but this is very dependent upon the type of food available. Commercially, they are mass produced using a factitious (alternative) food mite, *Carpoglyphus lactis*, which is mainly found on dried fruit, such as dried figs, prunes, raisins, etc., and on the debris in honeybee hives. Due to centuries of international trade in dried fruit, *C. lactis* is considered to be a cosmopolitan species that reduces the regulatory restrictions imposed on non-native species. Also, they are not recorded as causing damage to crops.

- *Crop/pest associations:* The mites feed on small arthropod prey such as whitefly eggs and young larval stages, thrips, and spider mites by piercing with their mouthparts and draining them of their contents. They also feed on grains of pollen. Like *A. andersoni* and *A. cucumeris,* these mites can be released in paper sachets that help ensure good establishment on most crops before the prey is present. Crops without pollen will result in slower establishment, which may require repeated introduction. However, the use of *C. lactis* in the rearing system improves predatory mite survival on most crops. Research in Canada has shown *A. swirskii* feeding on eggs of some beneficial organisms such as *Phytoseiulus persimilis* and *Aphidoletes aphidimyza,* which may have a negative effect on control of other pests.

- *Influence of growing practices:* At prolonged 15°C (59°F) temperatures, there is hardly any development and mites may die, although they will survive a short (overnight) spell at 8°C (46°F) as long as higher temperatures ensue the following day. The lowest recommended temperature for commercial use of *A. swirskii* is 20°C (68°F) and, while 25°C–28°C (77°C–82.5°F) is the

Figure 4.34   Adult and nymph *Amblyseius swirskii.*

Figure 4.35   Predatory *Amblyseius swirskii* among whitefly eggs.

temperatures in excess of 25°C (77°F) for active reproduction. This mite is used for control of whitefly (Figure 4.35) and young thrips with suppression of spider mites on protected cucumber and sweet pepper in much of Europe. Releases were made into the United States in 1983 as part of a control program for citrus pests. Adequate temperature and a supply of pollen or extrafloral nectaries appear essential for adequate establishment on most crops.

It is commercially available in many countries, but a license for release may be required in some.

- *Life cycle:* Adult female mites lay single eggs onto leaf hairs. The mites go through three immature stages—larva, protonymph, and deutonymph—before becoming adults. At 25°C (77°F), the entire cycle from egg to adult can optimum, they will survive at 40°C+ (104°F+). Humidity above 70% is also essential to prevent dehydration of eggs and larvae. Best results with this species have been seen on crops with mixed pest species such as thrips and whitefly, as frequently found on peppers. Some of the newer generation pesticides can be integrated with these mites but it is always advisable to check their use with the supplier.

PHYTOSEIULUS PERSIMILIS

The predatory mite *Phytoseiulus persimilis* is the principal biological control agent for the glasshouse spider mite *Tetranychus urticae*. It is easily differentiated from its host by its pear-shaped body and rapid movement. All but the larval stage are predatory and they will eat all stages of spider mite.

They are commercially available in many countries.

- *Life cycle:* Adult females lay up to five eggs per day, depositing them close to the host. Eggs are oval and almost twice the size of spider mite eggs (spider mite eggs are perfectly spherical) (Figure 4.36). On hatching, the six-legged larva moves little and does not feed but soon changes to a protonymph, which then starts to feed. Unlike spider mites, *P. persimilis* has no resting stages during its nymphal development. After the deutonymph

stage comes the adult, which normally mates within a few hours of development and may mate several times. Unmated females do not produce eggs.

- *Crop/pest associations: P. persimilis* was accidentally introduced to Germany on a shipment of orchids from Chile in 1958. Since then it has been used in all European countries and is now used worldwide. It is used wherever *T. urticae* are found (Figure 4.37), although its activity can be hampered on particularly smooth-leaved plants, such as carnation, or on leaves with sticky hairs (trichomes) (e.g., tomato). Tomato is a particularly difficult plant for *P. persimilis* activity as the stem is also covered in sticky hairs, reducing its mobility. Distribution of the mite is improved when leaves touch. On some crops where leaf contact is minimal, netting or even string may be used to provide walkways.

- *Influence of growing practices:* Temperature is possibly the most important factor for *P. persimilis* activity, there being little below 12°C (54°F); above 35°C (95°F) the mite ceases feeding and moves down the plant to cooler, shady areas. However, at 20°C–22°C (68°F–72°F) the mite's development time from egg to adult is twice as fast as that of the spider mite.

Figure 4.36   Adult female *Phytoseiulus persimilis* with her egg in front and spider mite egg behind.

Figure 4.37   *Phytoseiulus persimilis* nymph attacking glasshouse or two-spotted spider mite nymph (*Tetranychus urticae*).

Maximum egg production is between 17°C and 28°C (63°F and 82°F) and, as the normal male-to-female ratio is 1:4, it can be seen why this mite is so successful as a predator. Very high humidity (above 90%) is detrimental and also below 55% it can markedly disrupt egg survival.

TYPHLODROMUS PYRI

Typhlodromus pyri is a small, fast moving predatory mite found on a wide range of trees, often on the undersides of leaves close to the midvein. Adult females are pale (Figure 4.38) with a narrow, parallel-sided body, approximately 0.3 mm in length, males are slightly smaller. Eggs are translucent and oval shaped, approximately 0.16 × 0.11 mm. Larvae hatch but do not feed before developing into protonymphs, which resemble the adults in their shape.

- *Life cycle:* Females overwinter in crevices in the bark and emerge in the spring when they are found at high densities on the young leaves of vines and fruit trees. Eggs are usually laid singly on the undersides of leaves, often close to the midvein. Development to adult stage (Figure 4.39) takes about 2 weeks at 22°C (72°F). There are four or more generations per year depending on climatic conditions.
- *Crop/pest associations:* In temperate climates *T. pyri* is an important predator of spider mites and rust mites in apple (Figure 4.40) and pear orchards, as well as in vineyards. In warmer regions

Figure 4.38 Adult female *Typhlodromus pyri*.

Figure 4.39 *Typhlodromus pyri* mating pair.

Figure 4.40 Predatory *Typhlodromus pyri* attacking fruit tree spider mite (*Panonychus ulmi*) nymph.

different Phytoseiid species such as *Amblyseius abberans* and *Amblyseius californicus* occupy a similar niche. When there is no prey present, *T. pyri* can also feed on pollen, reaching very high population levels by this means. Mites are rarely found on leaves exposed to bright sunlight and, during the season, tend to concentrate within the midcanopy.

- *Influence of growing practices:* With careful use of only selective fungicides and insecticides, predatory mites can provide season-long control of spider mites and prevent any need for acaricide use. Like most predatory Phytoseiidae, *T. pyri* is very sensitive to pyrethroid insecticides and, if predatory mites

SUBCLASS: ACARI (PREDATORY MITES)

are to be encouraged, use of such insecticides should be avoided. Some populations of *T. pyri*, particularly in southern Germany, have developed resistance to a wide range of pesticides, particularly organophosphates and dithiocarbamates. Naturally occurring populations can be augmented by the release of commercially reared mites.

Family: Laelapidae (Soil-Dwelling Predatory Mites)

HYPOASPIS ACULEIFER AND *HYPOASPIS MILES*

Hypoaspis spp. are small, soil-dwelling mites inhabiting the top few centimeters of compost. They are also recorded from stored products, such as spoiled flour, rodent, and occasionally bird nests where they feed on animal parasit-oid mites. The mites are predatory, feeding on sciarid larvae (Figure 4.41) and other insects or mites. Female mites are the largest life stage, being up to 0.7 mm long and, like males, pos-sess a pale brown dorsal shield (Figure 4.42); immature mites are white in color.

These species are commercially available in many countries.

- *Life cycle:* The adult female deposits her oval white eggs on the soil surface and among the top layers of compost. These hatch into a six-legged larva, which passes through two eight-legged nymphal stages to reach adulthood some 18 days later at 20°C (68°F).

Figure 4.42   Predatory mite (*Hypoaspis miles*); note pointed end-to-dorsal shield.

Figure 4.43   Predatory mite (*Hypoaspis aculeifer*); note rounded end-to-dorsal shield.

- *Crop/pest associations:* Research has shown *H. miles* to be very efficient in controlling sciarid fly larvae in a range of composts and growing conditions, including mushroom houses. Other work has indicated a potential for control of thrips larvae, particularly when they pupate on the ground. This species is also widely used in interior atriums to control mealybug on low growing plants and hiding in structural cracks and crevasses (rocks, containers, etc.). *H. aculeifer* (Figure 4.43) has better activity against bulb scale mites (*Rhizoglyphus* spp.), making this species preferable for commercial use against these pests. Both species have shown a high tolerance to starvation as they can survive for 6–8 weeks in the absence of food, although water is required. When a mite captures its prey, it inserts its

Figure 4.41   Predatory mite (*Hypoaspis miles*) attacking sciarid fly larva.

saw-like mouthparts, which slice the internal tissues. These are then sucked up to leave a shriveled cadaver.

- *Influence of growing practices:* Hypoaspis spp. are commercially available and are an important adjunct to the biological armory where the control of soil pests requires nonchemical control measures. Temperatures below 11°C (52°F) cause inactivity and a cessation in egg hatch, whereas activity remains high at up to 30°C (86°F). These predatory mites show good tolerance to most leaf-applied pesticides.

Figure 4.44   Predatory mite *Machrocheles robustulus.*

Family: Macrochelidae (Predatory Mites)

MACROCHELES ROBUSTULUS

This family of predatory mites is found throughout the world, where they are associated with organic matter including animal dung, leaf litter, nests of birds, and nests of small mammals where associated prey is plentiful. Many of these mites occupy specialized and often transient habitats that attract several flying beetles and flies. It is on these larger insects that many *Macrocheles* mites are transported (phoretically) from site to site, often feeding on the eggs and young larvae of the insect that carried them in the first place. *Macrocheles robustulus* (Figures 4.44 and 4.45) is a cosmopolitan species that is now in commercial production in many countries for use against sciarid fly larvae and the soil stages of thrips. At approximately 0.7 to 0.8 mm in length, it is slightly larger than *Hypoaspis* spp. and is said to be a more voracious predator.

- *Life cycle:* Translucent eggs are deposited on and just below the ground surface. These hatch into minute white six-legged larva that, at 20°C (68°F), pass through two eight-legged nymphal stages to reach adulthood some 15 days later. Development is slow below 15°C (59°F) but they remain active up to 35°C (95°F).

Figure 4.45   Predatory mite *Machrocheles robustulus.*

- *Crop/pest associations:* These mites are commonly found in many agricultural and horticultural environments where they are part of the background fauna. They are now in commercial production and being used to help control several economically damaging pests. These include sciarid larvae and thrips pupae in compost. They may also feed on other soil dwelling organisms. However, they can become cannibalistic when prey levels are low. Recent research in the Netherlands showed that *M. robustulus* is more effective in controlling sciarids than *Hypoaspis* spp.

- *Influence of growing practices:* These mites prefer to live under the cover of soil/compost, with loose seeding media being far more suitable than bare, exposed soil. A more voracious predator than *Hypoaspis* spp., they may be used in a more curative role against sciarid outbreaks.

CLASS: CHILOPODA

Family: Geophilidae and Lithobiidae (Centipedes)

GEOPHILUS FLAVUS, STIGMATOGASTER SUBTERRANEUS, AND LITHOBIUS FORFICATUS

Centipedes, or hundred-legged worms, are not insects (Figure 4.46). They have a flattened, elongated body, with many segments, each bearing two legs (except the first behind the head and the last two). They are predatory against a wide range of soil-inhabiting organisms.

Stigmatogaster subterraneus and *Lithobius forficatus* are known as garden centipedes; once disturbed, they can be found scurrying around in moist areas. All centipedes have a distinct head that carries a pair of long, slender antennae made up of many joints, which may reach over one-third of the body length, as in *L. forficatus* (Figures 4.47 and 4.48). The first segment behind the head bears a pair of sharp claws containing poison glands that are used to seize and kill the prey. In some species the "bite" is harmful to humans but rarely lethal. The common *L. forficatus*, at 3 cm in length, can inflict a painful bite if handled roughly.

Centipedes of the Geophilidae family are longer (up to 12 cm) and snake- or worm-like, with a more rounded body, short antennae, and no eyes. They also possess glands used to spin a secretion that binds

Figure 4.47   Head view of brown centipede (*Lithobius forficatus*).

Figure 4.48   European brown centipede (*Lithobius forficatus*).

Figure 4.46   House centipede (*Scutigera coleoptrata*).

Figure 4.49   Snake centipede (*Geophilus flavus*).

eggs and spermatozoa together. Some species of *Geophilus* have over 150 legs (Figure 4.49).

MILLIPEDES

Millipedes, or thousand-legged worms, are also not insects (see Chapter 3, Figures 3.18–3.20). They have a more rounded body with four legs on each segment, do not possess poison glands, and generally have short antennae. They are plant feeders and scavengers of other food sources. An irritating and persistent odorous substance may be secreted as a defense against attack.

Figure 4.50   Garden centipede (*Stigmatogaster subterraneus*).

- *Life cycle:* Centipedes are relatively long lived, with some species known to live for up to 6 years. They overwinter as adults in dark humid areas such as compost heaps and leaf litter, under rotting timber, and other protected places. Egg laying begins in the spring and may continue through the summer months. Female centipedes often remain with their eggs to prevent them from becoming desiccated or from fungal attack if too moist. They may be seen licking the eggs, a process thought to involve the secretion of a fungicidal substance as well as removing excess moisture. Eggs usually hatch after a week or so into an eight-legged immature centipede and, because there is no larval stage, grow by adding more segments posteriorly, each with a pair of legs. Young centipedes may be confused with beetle larvae, particularly rove and carabid beetles (see p. 129). However, a quick leg count will easily identify one from the other. Insect larvae either possess three pairs of legs or none at all.
- *Crop/pest associations:* These predators are active hunters of many soil-dwelling organisms, including earthworms, millipedes, young slugs and snails, spiders, symphilids, woodlice, insect eggs, and beetle grubs. *H. subterraneus* (Figure 4.50) may also feed on plant roots, but generally does not cause economic damage. Their generally low numbers mean less competition between each other and is attributed to their longevity and ability to search for prey over a large area. They can also be cannibalistic and, in their juvenile stage, fall victim to other predators as well as to older centipedes.
- *Influence of growing practices:* Although common in most gardens, their need for a moist, dark environment means they are generally not found in intensive agriculture. Similarly, places where the soil regularly dries to a depth of more than a few centimeters are not conducive to the presence of centipedes.

CLASS: INSECTA (INSECTS)

Order: Odonata

SUBORDER: ANISOPTERA (DRAGONFLIES);
SUBORDER: ZYGOPTERA (DAMSELFLIES)

The term "dragonflies" loosely covers both the damselflies, which tend to be smaller, as well as the more robust Anisoptera, or true dragonflies. Damselflies have two pairs of similar wings that can be folded over their backs when at rest (Figure 4.51) and their bulbous eyes are always separated (Figure 4.52), usually with a bar of similar size or wider, than an individual eye. Dragonflies also have two pairs of wings, but the hindwings are broader than the forewings. When at rest, the wings are held at right angles to the body (Figure 4.53). The eyes of dragonflies are large and multifaceted, with up to 30,000 lenses to each eye (Figure 4.54), giving them all-round vision that can detect movement up to 10 m away. The eyes of true dragonflies are much closer together and may appear to be touching with just a fine line separating the two. Dragonflies are some of the fastest and most acrobatic fliers in the insect world. They can hover, fly backward, and even loop the loop; however, most cannot walk well. Damselflies are weaker fliers and usually remain close to water, whereas true dragonflies can be found several kilometers away from the nearest water. Dragonflies require permanent water for survival and may be found in most unpolluted wetland habitats

Figure 4.51   Newly emerged azure damselfly (*Coenagrion puella*) on an iris leaf with its exuvia (nymph case); note the wings folded back along the body length.

Figure 4.53   Adult female southern hawker dragonfly (*Aeshna cyanea*); note open wings.

Figure 4.52   Head view of azure damselfly (*Coenagrion puella*) showing separate compound eyes on either side of head and triangle of smaller ocelli in the middle.

Figure 4.54   Head of southern hawker dragonfly (*Aeshna cyanea*) showing large compound eyes joined in the middle and out-stretched wings.

including canals, ditches, and fresh water rivers, although some species prefer slightly acidic heathland bogs. Their other requirement is a source of submerged and emergent vegetation, as well as insect prey on which to predate.

Dragonfly nymphs have six legs, wing sheaths, and a hinged jaw that rapidly shoots out to catch passing prey. Damselflies have a narrower body with three leaf-like gills at the hind end, while true dragonflies have a broader body terminating with five spike-like appendages. Most dragonfly nymphs are ambush predators feeding on any organism that is smaller than they are. Their range of prey includes insect larvae, small fish, snails, and tadpoles and they can also be cannibalistic. Bottom-living species are covered in fine hairs that collect organic debris and silt. The hairs, along with their longer legs and antennae, sense passing prey while those known as hawkers have large eyes and hunt by sight closer to the water surface.

Figure 4.55 A mating pair of large red damselflies (*Pyrrhosoma nymphula*) showing the male holding the female by the neck.

- *Life cycle:* Dragonflies can be extremely territorial and, once a suitable stretch of water has been claimed, the male transfers sperm before mating from abdominal segment 9 to special accessory genitalia organs on segments 2 and 3. Using a pair of anal claspers, the male holds the receptive female by the neck (Figure 4.55). To avoid mating with insects of the wrong species, the claspers and neck groove fit together with lock-and-key precision. The pair can fly in tandem with the male dragging the female beneath. If she is receptive, her abdomen will rise to touch his accessory genitalia and they can continue flying in the wheel position. Mating like this can last from a few seconds to a few minutes and even over an hour with some species. During prolonged mating sessions, the pair is at greater risk from predators such as birds and will often fly among denser foliage or up into trees for protection.

Eggs can be laid singly or in batches over a period of several days or even weeks.

- Damselflies and hawker dragonflies possess a scythe-like ovipositor and inject an elongated egg into plant stems, leaves, rotting wood, or mud on or close to the water surface. Other species deposit spherical eggs in a jelly-like protective substance just below the water surface. Depending on species and temperatures, eggs may hatch either after 2 to 5 weeks or overwinter to hatch the following spring. The first-stage nymph is tadpole like and rapidly swims to find protection among vegetation, hollows in rocks or wood, and in mud. The nymph stages (Figures 4.56 and 4.57) may take only a few months in warmer climates, but the majority take 1 to 3 or more years and they can molt up to 15 times before climbing up and out from the water to emerge as an adult.

Figure 4.56   Young damselfly nymph.

Figure 4.58   Vacated nymph case or exuvia of a southern hawker dragonfly (*Aeshna cyanea*) above water line after adult emergence.

Figure 4.57   Southern hawker dragonfly (*Aeshna cyanea*) nymph on pond weed.

Dragonfly development is known as an incomplete metamorphosis as they have no pupal stage (Figure 4.58).

• *Crop/pest associations:* Dragonflies are mostly day-flying insects and use their excellent eyesight to locate their prey. The head is able to turn almost 360° on a slender neck, giving the insects all-round vision. Adults feed on almost any flying insect, including small flies, midges, and mosquitoes that are consumed on the wing. Some of the larger species can kill butterflies and moths as well as other dragonflies. In flight their legs form a basket for scooping up flying prey, which is passed forward to the serrated mandibles for eating. Most damselflies feed while resting on vegetation. Nymphs feed on insect larvae, water living crustaceans, leeches, slugs, snails, tadpoles, and small fish (Figure 4.59). If any limbs are lost during the nymphal stages, they may be replaced at the next molt.

• *Influence of growing practices:* Dragonflies require healthy, unpolluted

Figure 4.59   Southern hawker nymph (*Aeshna cyanea*) feeding on slug.

water and most permanent stretches of water are suitable. However, leaching of nitrates from farm land can lead to excessive algal growth, which reduces the available oxygen for nymphs and their food. Pesticides, oil, and other spillages can also drastically influence their survival. Though dragonflies are predators, they may become the food of many other predators such as birds, lizards, frogs, spiders, fish, water bugs, etc.

ORDER: DERMAPTERA

Family: Forficulidae (Earwigs)
FORFICULA AURICULARIA
Earwigs are found throughout the world. There are several theories concerning the origin of the English name. Some believe that it crawls into ears, and it is true that

earwigs do search out narrow cracks and passages coming to rest with their dorsal and ventral surfaces in contact with other surfaces. However, they only rarely occur in ears. A more likely explanation for the name is that it is derived from "ear-wing"—the delicate vein pattern of the hindwings—when unfolded, they resemble the human outer ear (the pinna) with its cartilaginous ridges and furrows. The order name Dermaptera, to which earwigs belong, gives a clue to the second way of describing the forewings, as it means "skin wing." The wings strikingly resemble flakes of human epidermis that peel off a few days after severe sunburn.

Many species are omnivorous, feeding as predators, scavengers, and, occasionally, primary plant pests. Adult females are protective of their brood and in some species carry food to the young. They may be considered to be social insects. The European earwig (*Forficula auricularia*) (Figures 4.60–4.63) is an efficient predator

Figure 4.60   Adult male earwig (*Forficula auricularia*).

Figure 4.61 **Adult female earwig (*Forficula auricularia*) on apple leaf.**

Figure 4.63 **Male earwig nymph (*Forficula auricularia*) on apple leaf.**

of aphids and insect eggs. It is able to survive with low densities of prey in many situations, including in cereal, hop, and orchard crops. Adults are around 20 mm long with a flattened reddish-brown body. The tip of the abdomen has a pair of pincers that are curved in the male and almost straight in the female. The veinless forewings (elytra) are short and serve to protect the extremely delicate folded hindwings. The insect can fly, but this is rarely observed. Earwig nymphs resemble adults but have wing buds, not wings; these become larger at each nymphal molt.

- *Life cycle:* Adults and eggs overwinter and are found below ground in moist, sheltered places. The adults form cells in the soil and the nymphs are protected by the parents until they reach the third instar. Eggs hatch in early spring and the young nymphs feed on young leaves and other plants. Mature adults first appear in early summer. There is one generation each year, although overwintering females lay eggs that hatch later than those laid in the autumn to produce a winter and spring brood. Earwigs are mainly nocturnal, hiding within flowers, leaf sheaths, and hollow plant stems; under bark; or among dense mats of vegetation close to the soil surface unless disturbed. Their climbing ability

Figure 4.62 **Adult male earwig (*Forficula auricularia*) with cerci raised in defensive reaction.**

contributes to their role as predators of pests such as aphids and has led to the gardening trap of an upturned flower pot on a stick with crumpled paper in the pot. The pot should be in contact with the upper plant canopy while the paper provides the surface contact the insect needs, thus encouraging these predators to remain close to the aphids. The same technique can be used to collect earwigs for removal from particularly delicate flower blooms such as exhibition dahlia heads.

- *Crop/pest associations:* Earwigs are mainly scavengers on decaying plant tissue, but they also feed on green plant matter and living and dead insects. Growers recognize them as a pest, particularly when they damage flowers and leaves. They also chew fruits close to the ground, including strawberries or ripe black currants. In North America they have been the subject of a biological control program. However, they can be important predators of pests such as aphids, especially in crops of top fruit and hops, where aphids are a major problem; earwigs can outclimb most other predatory groups. In arable land they also eat aphids, but their close association with field boundaries persists through the summer, limiting their biocontrol potential compared with more dispersive insects such as the carabid beetles.

- *Influence of growing practices:* Earwigs are susceptible to insecticides and to the disruption of their field boundary habitat by cultivation and the removal of perennial grasses (see p. 13). Monitoring of earwigs can be done by using corrugated cardboard traps attached to tree trunks or wooden poles. They are also parasitized by various tachinid flies (see p. 146).

ORDER: HEMIPTERA

Suborder: Heteroptera (Predatory Bugs)

Heteropteran bugs are so named because their forewings have different forms from those of the hind wings. The forewings have a hard, leathery basal area and membranous tip, and the hind wings are wholly membranous. The antennae are frequently long in relation to the size of the body but in nearly all species have only four obvious segments. These insects are all noticeable for having tubular piercing mouthparts (Figure 4.64) and many are plant pests, often causing severe damage. However, several are very active predators feeding on a wide range of organisms. There are many aquatic predatory species like water boatmen (*Notonecta glauca*), pond skaters (*Gerris lacustrus*), and several that have specialized as blood feeders, generally on other insects. Distinguishing between an herbivorous capsid bug and a predatory black-kneed capsid is not easy.

Many heteropteran predators include plant juices in their diet and most lay their eggs into plant tissue, often leaving a small, raised blister-like opercula visible on the leaf surface. After depositing eggs, adults frequently feed from the plant wound. There are many predatory species from several families and these attack a wide range of prey species—often as plant or habitat

Figure 4.64   Adult *Podisus maculiventris* with stylet inserted in a tomato moth caterpillar (*Lacanobia oleracea*).

specialists (Figures 4.65 and 4.66). Feeding begins when the grooved rostrum or proboscis selects a suitable site for the mandibular stylets within to penetrate, rasp, and cut the prey tissue before the maxillary stylets enter the wound. These delicate needle-like tubes are protected within the mandibular stylets and form canals that deliver saliva

Figure 4.67   Adult *Anthocoris* spp. feeding on flower pollen and nectar; the name comes from the Greek *anthos*, for flower, and *koris*, for bug.

Figure 4.65   Adult flower bug (*Orius niger*) feeding on peach–potato aphid (*Myzus persicae*).

containing proteolytic enzymes. Predigested, liquefied prey contents are removed. However, in the absence of insect or mite prey, they are able to survive on plant sap and pollen (Figure 4.67). Many are able to penetrate human skin to give a small "bite."

Family: Gerridae (Pond Skaters)

COMMON POND SKATER OR COMMON WATER STRIDER (GERRIS LACUSTRIS)

Common pond skaters are long, narrow insects, 15 to 20 mm in length, usually dark brown or gray with wings folded tightly along the body. Their eyes protrude from the side of the head. Females are usually larger than males. Legs and the underside of the body are covered with fine, water-repellent hairs allowing them to float on the water's surface tension (Figures 4.68 and 4.69). The hairs also sense vibrations of falling or passing prey, and their struggling, which produces small ripples, guides the predator toward its meal. The front legs are used to catch and manipulate prey while the middle legs propel and the rear pair act as rudders to steer the insect. Being true bugs, they have sharp, piercing mouthparts that hold the prey and suck out the juices. They may give a painful bite if handled.

- *Life cycle:* Males use their front legs to hold a female for mating while both float on the water surface. Eggs are laid on foliage around the edge of ponds or

Figure 4.66   Adult *Psallus* sp. feeding on green apple aphid (*Aphis pomi*).

on still water. Emergent nymphs drop into the water and swim to the bottom, where they feed on almost any suitable organism that is smaller than themselves (Figure 4.70). Pond skaters are mainly found from midspring to early autumn. From late autumn they fly to find shelter to overwinter, sometimes quite a distance from water and commonly in small groups.

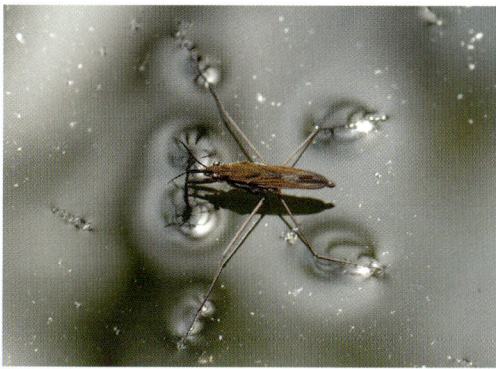

Figure 4.68   Adult pond skater (*Gerris lacustris*).

Figure 4.69   Side view of adult pond skater (*Gerris lacustris*).

Figure 4.70   Young nymph of *Gerris lacustris*.

- *Crop/pest associations:* Although pond skaters do not directly feed on plant pests, they will catch and feed on a wide range of insects that accidentally fall onto the water. They also feed on mosquito larvae and even their own nymphs. They are extremely agile and can jump off the water surface to fly away from potential predators and catch small flying insects.
- *Influence of growing practices:* Pond skaters require healthy, unpolluted water that is usually static or slowly moving. As with other water-living organisms, excess nitrates leaching into the water can lead to excessive algae growth and a slow stagnation, reducing the water's capacity to generate more complex life. Excess use of slug pellets, particularly metaldehyde-based ones, can also leach the active ingredient into water, killing many water-living insects. Pond skaters may become the food of many other predators such as birds, lizards, frogs, spiders, fish, dragonflies, etc.

Family: Anthocoridae (Predatory Bugs/Flower Bugs)

ANTHOCORIS NEMORALIS

Nymph and adult *Anthocoris nemoralis* are active predators of aphids (Figure 4.71), spider mites, and psyllids; they may also feed

Figure 4.71   Adult *Anthocoris nemoralis* feeding among colony of black cherry aphid (*Myzus cerasi*).

Figure 4.72   Adult flower bug (*Anthocoris nemoralis*) sucking juice from raspberry fruit.

Figure 4.73   *A. nemoralis* nymph feeding on larva of apple leaf curling midge (*Dasineura mali*).

on plant sap and ripe fruit (Figure 4.72). Common throughout Europe on wild and cultivated shrubs, bushes, and trees, *A. nemoralis* may become very abundant on commercially managed or neglected fruit trees. Neglected orchards typically contain very high numbers of *A. nemoralis* for 1 or 2 years after being abandoned, after which time their numbers decline rapidly. Adult *A. nemoralis* are distinguished from *A. nemorum* (see following section) by the coloration of the second and third antennal segments. Antennae are shorter than those of *A. nemorum*.

- *Life cycle:* There are two generations per year. Adults overwinter beneath bark, among dead leaves, and in hedgerows. Females emerge from hibernation and lay eggs from spring to early summer. Eggs are inserted into the leaf near the midvein, often in groups. Freshly laid eggs are white, turning reddish-brown shortly before the emergence of the nymph. The first instar nymph is clear or yellow in color. Subsequent instars are reddish brown in color (Figure 4.73).
- *Crop/pest associations: A. nemoralis* is known as a predator of fruit tree red spider mite in apple orchards, but also attacks aphids and psyllids in many crops. It tends to prefer small prey items.
- *Influence of growing practices:* Increased plant diversity, particularly with flowering plants in and adjacent to

orchards, can provide alternative food sources and refugia as well as suitable overwintering sites. Anthocorid bugs (*Anthocoris* and *Orius* spp.) are usually less sensitive to pesticides than mirid bugs (see Table 4.1, pp. 66–69).

ANTHOCORIS NEMORUM

Anthocoris nemorum is similar to *A. nemoralis* in appearance and in habits, but the antennae are longer than those of *A. nemoralis*. Dark parts of the forewings are matte in appearance.

- *Life cycle:* There are two or three generations per year. Adults overwinter beneath bark, among dead leaves, and in hedgerows. Females emerge from hibernation and lay eggs from spring to early summer. Females lay about two eggs per day; these are inserted into the leaf near the midvein, often in groups up to eight. There are five nymphal stages and the first instar is clear or pale yellow. Subsequent instars are reddish-brown in color (Figure 4.74).
- *Crop/pest associations: A. nemorum* is a common, naturally occurring predator on trees and shrubs. In fruit orchards it attacks a wide range of insect and mite prey and has been observed feeding on the sessile larvae of the predatory ladybird *Stethorus punctillum*. *A. nemorum* feeds by piercing its prey

Figure 4.74   Adult *Anthocoris nemorum* feeding on green apple aphid (*Aphis pomi*); note aphid excreting droplets of alarm pheromone to warn other aphids to flee.

Figure 4.75   *A. nemorum* nymph feeding on black cherry aphid (*Myzus cerasi*).

from the side with its short rostrum and then sucking out the body contents (Figure 4.75). Adults consume on average 50 spider mites per day.

- *Influence of growing practices:* The presence of flowering plants under trees and in orchard margins can provide alternative prey and lead to increased numbers of these predators. *A. nemorum* is commonly found on blackberries in the autumn. Brambles and hedgerow plants can be important overwintering sites for *A. nemorum*.

ORIUS LAEVIGATUS

Orius laevigatus is a common and important predator of aphids, midge larvae (Figure 4.76),

spider mites, and thrips (Figures 4.77 and 4.78) in orchards and in glasshouse vegetable crops. It is similar to *O. majusculus*, but smaller (2.2–2.5 mm in length), and also to several other *Orius* species (*O. minutus*,

Figure 4.76   Adult *Orius laevigatus* feeding on blackberry midge (*Dasineura plicatrix*).

Figure 4.77   Adult *Orius laevigatus* feeding on Western flower thrips larva (*Frankliniella occidentalis*).

Figure 4.78   *Orius laevigatus* nymph feeding on Western flower thrips larvae (*Frankliniella occidentalis*).

Figure 4.79   Head view of adult *Orius laevigatus* feeding on plant sap; note red compound eye.

O. *vicinus*, O. *horvathi*). It is necessary to examine the genitalia to identify species occurring in the wild. *Orius* spp. are commonly called flower bugs as they are usually found close to or within flowers, where they feed on insects and mites; they may also feed on pollen and sap (Figure 4.79).

They are commercially available in many countries.

- *Life cycle:* Two generations occur per year in northern Europe and three per year in Mediterranean regions. Eggs are laid in groups of three to four in the midvein on the undersides of leaves. Eggs hatch from spring onward; nymphs and adults are found continuously until autumn.
- *Crop/pest associations:* Adults and nymphs are voracious predators and are commercially available as biological control agents all year round. When prey is scarce, O. *laevigatus* has been known to feed on predatory mites.
- *Influence of growing practices:* O. *laevigatus* is commercially available as a biological control agent and is sold primarily for use in protected crops. O. *laevigatus* can feed on lepidopteran eggs when alternative food is scarce.

ORIUS MAJUSCULUS
Orius majusculus is a common and important predator of aphids, thrips, and spider

mites in neglected and commercial orchards. Adults are small (2.6–3.8 mm in length) with flattened, rounded bodies (Figure 4.80) very similar in appearance to O. *minutus*.

The species are commercially available in many countries, where they are introduced to glasshouses and other commercial crops.

- *Life cycle:* Females emerge from hibernation through the spring months. Eggs are laid in groups of three to four in the midvein on the undersides of leaves. Eggs hatch from spring onward, and nymphs and adults are found continuously until autumn. The time from oviposition to adult emergence varies from 24 to 40 days. There are typically two or three generations per year.
- *Crop/pest associations:* Adults and nymphs are active predators of mites and small insects (Figures 4.81 and 4.82). If fruit tree red spider mites are present in high numbers, they will be the preferred prey with all stages being attacked. If prey is scarce, they are cannibalistic. Adult O. *majusculus* consume about 35 adult mites per day and nymphs consume about 22.
- *Influence on growing practices:* Eggs are generally protected in the leaf midvein. Where insecticides are used, those that

Figure 4.80   Adult *Orius majusculus* feeding on Western flower thrips (*Frankliniella occidentalis*).

Figure 4.81   *Orius majusculus* nymph feeding on Western flower thrips larva (*Frankliniella occidentalis*) larva.

Figure 4.82   Late instar nymph of *Orius majusculus* feeding on thrips larva; note developing wing buds.

are active for a short period are less damaging than persistent products. The use of microbial insecticides such as *Bacillus thuringiensis* reduces the risk to many predators. The presence of patches of wild flowering plants, either under the trees or in a part of the orchard, provides a refuge enabling recolonization when toxic sprays are used and can provide an alternative source of food when prey is scarce on the trees.

Family: Miridae (Predatory Bugs)

ATRACTOTOMUS MALI

Atractotomus mali is a small predatory insect found on hawthorn and apple trees and, more frequently, in neglected orchards. Adults have a characteristic appearance and

Figure 4.83   Adult *Atractotomus mali* searching for prey on apple leaf.

are black or reddish-brown with a covering of fine white hairs (Figure 4.83). Males are 3.0–3.5 mm long, whereas females are 3.3–4.0 mm long. In nymphs and adults the second antennal segment is enlarged and is dark, while the third and fourth antennal segments are pale and slender.

- *Life cycle:* One generation occurs per year. Eggs are laid in bark, often in clusters where they overwinter, hatching in the spring. Nymphs are active from early to midsummer and adults begin to appear from midsummer onward.
- *Crop/pest associations:* A. *mali* attacks spider mites, other mites, and occasionally aphids and moth larvae in fruit crops throughout Europe. It can also be partly phytophagous if insect or mite prey is scarce.
- *Influence of growing practices:* A. *mali* prefers neglected or low-input orchards and is generally more abundant where there are established hedgerows close to an orchard.

BLEPHARIDOPTERUS ANGULATUS

Blepharidopterus angulatus is commonly known as the black-kneed capsid because of dark spots at the base of each tibia. Formerly one of the most common predators in apple orchards in southern England, *B. angulatus* has become scarce in recent years. It is a localized predator of spider mites and small insects on fruit trees, particularly apple, pear, plum, and cherry. It is also found on many other tree species. The dark spot is present in nymphs and adults. Males are 5.1–5.7 mm in length; females are 5.7–5.8 mm long.

- *Life cycle:* There is one generation per year. Eggs are laid deep into 1- or 2-year-old wood, usually singly. A characteristic bump in the wood develops around the egg about 3 days after laying. *B. angulatus* overwinters in the egg stage. Nymphs begin to emerge from early to midsummer and are yellow to green in color, with red eyes. There are five instars. Adults are bright green (Figure 4.84) with brown antennae; males are larger and darker than females.

- *Crop/pest associations: B. angulatus* is a voracious predator of *Panonychus ulmi* (fruit tree red spider mite), preferring to attack adult female mites. It takes about 3 minutes for an adult to suck dry an adult female spider mite. A single *B. angulatus* can consume as many as 4,300 mites during its lifetime, with as many as 3,000 consumed by the adult. Although primarily carnivorous, like many mirid bugs *B. angulatus* can feed on plant material, particularly rotten fruits on trees or on the ground.

Influence of growing practices: B. angulatus commonly lays eggs in the wood of *Alnus glutinosa* and *Betula alba*. Suitable tree species in margins of orchards may encourage oviposition. *B. angulatus* is highly sensitive to pesticides.

DERAEOCORIS RUBER

Adult *Deraeocoris ruber* are large (7.0 mm in length) and light brown or black in color with red flashes on the cuneus (wedge-shaped region at the end of the forewing) (Figure 4.85). Nymphs are crimson reddish with a characteristically wide abdomen bearing black spines.

- *Life cycle: D. ruber* has one generation a year. Eggs are laid singly in 1 or 2-year-old wood in late summer and hatch from early to midsummer the following year. Adults appear throughout the summer months and into early autumn.

- *Crop/pest associations:* Adults and nymphs are active predators of aphids, particularly the woolly apple aphid *Eriosoma lanigerum,* but also feed on psyllids and mites in apple and pear orchards. Nymphs (Figures 4.86 and 4.87) can occur in large numbers, but adults disperse and are rarely found at levels greater than five individuals per tree. *D. ruber* is almost entirely predatory and can attack other predators; it is also cannibalistic.

Figure 4.84   Adult *Blepharidopterus angulatus* on apple leaf. (Photograph courtesy of David V. Alford.)

Figure 4.85   Adult *Deraeocoris ruber* opening wings for flight. (Photograph courtesy of Peter Wilson/FLPA.)

Figure 4.86 Nymph of predatory bug *Deraeocoris ruber.*

Figure 4.87 Head view of *Deraeocoris ruber;* these polyphagous predators also feed on plant sap.

Figure 4.88 Adult *Heterotoma planicornis.*

- *Influence of growing practices:* D. ruber is more frequently found in weedy orchards, particularly if there are nettles present under the trees or in the margins. It is rarely found in orchards with bare-earth strips beneath trees.

HETEROTOMA PLANICORNIS

Heterotoma planicornis is commonly found in neglected orchards throughout Europe. Males are 4.6–5.3 mm in length; females are 4.9–5.5 mm. Adults are purple to dark red in color with a long thin body and characteristic enlarged and flattened second antennal segments (Figure 4.88). Nymphs are red in color and also have an enlarged second antennal segment (Figure 4.89).

- *Life cycle:* H. planicornis have one generation a year. Eggs are laid in late summer into young wood. Eggs hatch from late spring to midsummer the following year. Adults appear throughout the summer months and into early autumn.
- *Crop/pest associations:* Adults and nymphs are active predators of spider mites and aphids in apple and pear orchards. High numbers have been observed together with outbreaks of pear sucker (*Psylla pyricola*). H. planicornis is partly phytophagous and leaves and fruits may be attacked.
- *Influence of growing practices:* H. planicornis is more frequently found in orchards with an understory of mixed vegetation. Like *Deraeocoris ruber*, it appears to favor orchards where nettles are present, either under the trees or in the margins.

Figure 4.90   Adult *Macrolophus pygmaeus;* **note black first antennal segments.**

Figure 4.89   Predatory bug nymph (*Heterotoma planicornis***) searching for aphids on apple leaf.**

Figure 4.91   *Macrolophus pygmaeus* nymph among late instar glasshouse whitefly (*Trialeurodes vaporariorum***).**

MACROLOPHUS PYGMAEUS (SYN. *M. CALIGINOSUS*)

Macrolophus pygmaeus is a polyphagous predatory mirid occurring naturally in a wide range of crops throughout the Mediterranean region. Adults are green and 6.0 mm in length, with long legs and antennae that have a black first segment (Figure 4.90). Nymphs are green to yellowish-green (Figure 4.91) and are found mainly on the undersides of leaves. All mobile stages are predatory against a range of small invertebrates including glasshouse whitefly (*Trialeurodes vaporariorum*), (Figure 4.92) tobacco whitefly (*Bemisia tabaci*), aphids, thrips, moth eggs, and spider mites.

This species is commercially available in many countries, but a license for release is required in some.

- *Life cycle: M. pygmaeus* is mainly found on plants of the Compositae

and Solanaceae families in which adult females lay their eggs. The eggs are inserted deeply in the plant stems and leaves with a tiny breathing tube visible only by hand lens. Development is relatively slow (30 days at 25°C [77°F], increasing to 50 days at 20°C [68°F]), compared to the principal prey, whitefly (22 days at 25°C [77°F] and 30 days at 20°C [68°F]). However, *M. pygmaeus* are partly phytophagous, feeding on plant sap and pollen, which helps their early inoculative release on protected crops before pest levels begin to establish.

- *Crop/pest associations: M. pygmaeus* is commercially available as a glasshouse predator of whitefly and other pests. Some countries require a license for

Figure 4.92 *Macrolophus pygmaeus* **nymph cleaning stylet after feeding on glasshouse whitefly pupa (*Trialeurodes vaporariorum*).**

its release. Research has shown that greater than four times more eggs are produced when the predator has whitefly in its diet than when feeding on plant sap alone. Under ideal conditions of suitable host plant, plentiful food supply, and reasonably high temperatures, the predator can become a plant pest. This particularly applies to cherry tomatoes, where fruit loss has been observed. They are not recommended on gerbera, due to flower damage, or on cucumber and pepper due to poor establishment.

- *Influence of growing practices:* Additional pest control agents, such as *Encarsia formosa* for whitefly, are required during the early part of the season while *M. pygmaeus* establishes. Diapause in short day lengths does not seem to be a problem to their early season establishment, as females collected during midwinter in the south of France

had ripe eggs in their ovaries. Like all Miridae (see Table 4.1, pp. 66–69), the adults and nymphs are sensitive to most insecticides, while the eggs are protected within the leaf or stem. Short-persistence pesticides, such as natural pyrethrum, can be used to reduce predator numbers when crop damage is suspected without eliminating the bug altogether.

PILOPHORUS PERPLEXUS

At first sight these predatory bugs resemble ants, in both coloration and movement. The adult *Pilophorus perplexus* is 4.0–4.9 mm in length, dark brown in color with two conspicuous white bands across the forewings (Figure 4.93). Nymphs are also dark brown with a broad white spot at the base of the abdomen (Figure 4.94). Both adults and nymphs run rapidly up and down branches in search of prey. *P. perplexus* has been observed as an abundant predator on fruit trees where ants were also active.

- *Life cycle: P. perplexus* has one generation a year. Eggs are laid in late summer and overwinter, hatching in spring to early summer. Adults appear from mid- to late summer.
- *Crop/pest associations:* Adults and nymphs are active predators of aphids, particularly *Aphis pomi* on apples, but also feed voraciously on Lepidoptera larvae, psyllids, and mites

Figure 4.93 **Adult *Pilophorus perplexus*.**

Figure 4.94   Nymph of predatory bug and ant mimic *Pilophorus perplexus*.

Figure 4.95   Adult *Podisus maculiventris*; also known as spined soldier bug and stinkbug.

in fruit orchards. *P. perplexus* is also commonly found on oak and other deciduous trees.

- *Influence of growing practices:* *P. perplexus* is usually more common in orchards with established hedgerow margins and may also favor wild plants in uncropped refugia.

Family: Pentatomidae (Shield Bug)

PODISUS MACULIVENTRIS

Podisus maculiventris is a predatory shield bug found from Canada to Mexico and may be known as the spined soldier bug or stinkbug (an odor is released when it is attacked or squeezed). Adults are 9–13 mm long with prominent spines on the broad "shoulders" situated just behind the head (Figure 4.95). Several similar looking shield bugs are pests of plants and may be differentiated from *P. maculiventris* by the presence of a distinctive dark patch on the tip of each forewing. Adults and larvae possess a long

Figure 4.96   Young nymph of predatory bug *Podisus maculiventris* feeding on dipteran larva.

pointed proboscis to stab (Figure 4.96) and carry their prey. This is kept folded under the body when not in use for feeding.

This species is commercially available in many countries, but a license for release is required in some.

- *Life cycle:* Adult females can produce several hundred eggs at a time in groups of 10–30. These can be deposited on leaves or stems of plants, usually in close proximity to their prey. The eggs hatch after a few days to bright red and black nymphs that become more deeply colored as they mature. First instar nymphs remain together in a cluster after emerging and are reported not to feed on insect prey. Older nymphs are voracious feeders (Figure 4.97) and can be cannibalistic when prey is scarce.

- *Crop/pest associations:* Young nymphs may be found feeding together from the same prey, particularly if it is a large caterpillar. Larger *P. maculiventris* can be seen running around with a prey body harpooned on their proboscis and may feed from the same larva until most of the juices are consumed. Good control of Colorado potato beetle (*Leptinotarsa decemlineata*) has been achieved in several countries with *P. maculiventris*. Also, satisfactory control of caterpillars is reported on protected crops in Europe, where it is introduced under grower license (not permitted in UK).

- *Influence of growing practices:* This predator prefers a warm environment with temperatures above 18°C (64°F) for continual development. It requires a source of plant sap in its diet but is not reported as causing any significant plant damage during feeding. Its main host plant range includes cucurbits, aubergine, pepper, potato, and tomato. A polyphagous predator, it will feed on other beneficial organisms as well as on pest species.

Family: Psyllidae (Psyllids)

APHALARA ITADORI

This psyllid is a sap-sucking insect with great potential as a biological control agent against Japanese knotweed (*Fallopia japonica*) (Figures 4.98 and 4.99). Adults are approximately 2.5 mm in length with mottled shades of orange, red, and brown, but with a more solid dark brown band toward the wing tips (Figure 4.100). They are native to Japan and have a very narrow host plant range restricted to *F. japonica* and the similarly damaging hybrid Bohemian knotweed (*Fallopia* x *Bohemica*) with some ability to develop on giant knotweed (*F. sachalinensis*).

Figure 4.97 **Nymph of predatory bug *Podisus maculiventris* feeding on tomato moth caterpillar (*Lacanobia oleracea*).**

Figure 4.98 **Invasive flowering Japanese knotweed (*Fallopia japonica*).**

Figure 4.99   Regrowth of Japanese knotweed (*Fallopia japonica*); this invasive weed can grow from miniscule sections.

Figure 4.101   Nymph of *Aphalara itadori*. (Photograph courtesy of Dr. Richard [Dick] Shaw, CABI, UK.)

Figure 4.100   Adult *Aphalara itadori* with elongated creamy white eggs on leaf surface. (Photograph courtesy of Dr. Richard [Dick] Shaw, CABI, UK.)

- *Life cycle:* Adult females deposit elongated creamy white eggs, approximately 0.4 mm in length, flat to the plant surface. These hatch to minute nymphs, approximately 0.5 mm in length, that feed by piercing plant cells and sucking sap (Figure 4.101). There are five nymphal stages before adulthood is reached; the whole cycle from egg to adult takes 33 to 35 days at 23°C (74°F). Adults can live for up to 2 months and females can each lay over 650 eggs. Adults in their native range disperse in the autumn to overwinter on shelter plants, usually conifers including Japanese cedar (*Cryptomeria japonica*) and Japanese red pine (*Pinus densiflora*). They do not appear to feed significantly on their overwintering shelter plants as no feeding damage has been observed.

- *Plant associations:* Japanese knotweed (*F. japonica*) is one of the most troublesome invasive alien weeds in many parts of Europe and North America. It is a vigorous, herbaceous perennial with light green, hairless, tubular stems, often having reddish flecks. Mature plants can reach 3 to 4 m in length, terminating in branches of creamy white flowers. Leaves are broad oval in shape, up to 14 cm long by 12 cm wide. Stems and leaves die off after the first frosts and the plant survives winter as root material and a large crown (rhizome) from which new growth starts in the late spring. All motile stages of *A. itadori* feed by sap sucking and laboratory studies show they can cause significant damage to the host plant. Developing nymphs, particularly the fourth and fifth stages, do the most feeding.

- *Influence of growing practices:* Extensive research has been done to determine the full host plant range of this psyllid, including monitoring egg laying and survival through to adulthood. This has shown that of over 90 plant species tested, only 0.6% of eggs were laid on nontarget plants and none of these reached adulthood. These results (Shaw, Bryner, and Tanner 2009) indicate that

A. itadori offers great potential as a host-specific, classic biocontrol organism against the nonindigenous weed *F. japonica* and should cause stunting, reduced flowering, and reduced viability. It is likely that plants weakened by the insect would also be more susceptible to other control measures such as herbicides. Similarly, several herbicides can be integrated with the insect, leading to improved weed control.

Family: Thripidae (Predatory Thrips)

FRANKLINOTHRIPS VESPIFORMIS

Franklinothrips vespiformis are predatory thrips that feed on other thrips species. These are small elongated insects similar in shape to most plant-feeding pest species except that adults have a wasp-like waist (Figure 4.102),

Figure 4.102   Adult predatory thrips (*Franklinothrips vespiformis*) (black) next to Western flower thrips larva (*Frankliniella occidentalis*).

Figure 4.103   Larval stage of *Franklinothrips vespiformis* showing raised abdomen.

giving the appearance of being broader across the abdomen. Their larvae appear hump backed and have an upturned tail, similar to that of a scorpion (Figure 4.103). Both adults and larvae mimic ants with their rapid, often jerky movement. Adult males are rare, 2.5 mm in length, and uniformly brown in color; females are 3 mm long and mostly brown with yellow stripes across the abdomen and antenna. Larvae are orange-red to light brown in color. Various species of *Franklinothrips* occur naturally in most tropical and subtropical areas around the world.

This species is commercially available, but a license for release is required in some countries.

- *Life cycle:* Eggs are inserted in plant tissue, usually in close proximity to phytophagous thrips. These hatch to minute reddish larvae that feed on larvae, prepupae, and pupae as they mature (two larvae instars). They can also feed on adult thrips. Egg to adult takes approximately 25 days at 25°C (77°F). Females can produce 150 to 200 eggs over a 4- to 6-week period. Second instar larvae spin a silken cocoon, either on the underside of a leaf next to a vein or on the ground. Within the cocoon, pupation takes place, followed by adult emergence.
- *Crop/pest associations:* Their main commercial use is to control gregarious thrips such as *Parthenothrips dracaenae* (palm thrips) and *Echinothrips americanus* (numerous common names). These are more difficult to control with predatory mites. They will also feed on *Thrips palmi* (melon thrips, a notifiable pest in many countries) and *Frankliniella occidentalis* (Western flower thrips) (Figure 4.104). However, these predators have also been recorded feeding on spider mites, aphids, and whitefly as well as other small insect species.
- *Influence of growing practices:* Temperatures above 18°C (64°F) are

Figure 4.104   Second instar *Franklinothrips vespiformis* (orange/red) with Western flower thrips (*Frankliniella occidentalis*) prepupa.

Figure 4.105   Head of green lacewing larva (*Chrysopa perla*) searching for aphids.

required for activity but they can remain active at up to 40°C (104°F), making them ideal for commercial use in protected plant collections in botanic gardens, conservatories, interior landscapes, and nurseries as long as the minimum temperatures can be achieved. In tropical and subtropical zones, they may be used outside. As this is a tropical species, licenses may be required for release in some countries.

ORDER: NEUROPTERA

Adult neuropterans are either predatory or omnivorous, feeding on fungal hyphae, pollen, and honeydew. Although some species are specialist feeders, the larvae of most species are regarded as generalist predators, feeding with formidable looking mouthparts that both pierce and suck the body juices of prey (Figure 4.105). Adults of the genus *Chrysoperla* belong to the second group and feed only on honeydew and pollen. To overcome any dietary shortages of essential amino acids, they have a symbiotic yeast (*Torulopsis* spp.) living around their mouthparts.

The digestive system of a lacewing larva has no direct passage to the anus. Instead, the small amount of solid material ingested accumulates within the gut, which is emptied when it reaches adulthood. This is due to the mainly liquid diet of lacewing larvae.

Figure 4.106   Adult snake fly (*Raphidia* spp.), usually found in fruit orchards.

Figure 4.107   Adult brown lacewing (*Hemerobius humulinus*) on plum leaf.

Chrysoperla spp. are easier to mass rear than *Chrysopa* spp.; consequently, more research work has been done on the former genus. Adult *Chrysoperla* spp. can also be attracted to crops before pest numbers become excessive by spraying honeydew, sucrose solution, or molasses over the plants. There are also some less common lacewing adults (Figures 4.106 and 4.107).

Family: Chrysopidae (Predatory Lacewings)

CHRYSOPA PERLA

Similar to and often confused with *Chrysoperla carnea*, the adults of the *Chrysopa* genus are all predatory, feeding on a range of soft-bodied prey. *C. perla* adults are 15–20 mm in length with their wings folded and have a wing span of up to 30 mm (Figure 4.108). They are commonly found across northern Europe and, rarely, in the hotter areas of the Mediterranean region or much of Asia.

- *Life cycle:* Blue/green eggs are laid at the end of a mucus stalk up to 5 mm in length; they are deposited singly or several may be found close together, depending mainly on the quantity of available food for the larvae (Figure 4.109). Mature larvae spin a silken cocoon in which they pupate and from which they emerge as adults.

Figure 4.108   Adult green lacewing (*Chrysopa perla*) on pink aqualigia flower.

C. perla overwinter (diapause) as larvae in cocoons and continue development as conditions warm the following spring.

- *Crop/pest associations:* Although regarded as generalist predators, research has shown that a diet of certain aphids alone will allow full development through to adulthood, but can result in reduced viability of the adult insect. As with other lacewings, the adults prefer an open plant structure to fly through, but the larvae tend to be found hidden within a leaf canopy.

- *Influence of growing practices:* The mass production of *Chrysopa* spp. is more complicated than that of *Chrysoperla* spp.; consequently, most lacewings purchased for commercial use are *Chrysoperla carnea*. Nevertheless, *C. perla* are widely used in several countries alongside other beneficial organisms to control aphids on many crops. They may be introduced as young larvae or more commonly as eggs.

CHRYSOPERLA CARNEA (COMMON GREEN LACEWINGS)

Known as common green lacewings after the delicate wing venation of adults (Figure 4.110), or the aphid lion after the voracious appetite of its larvae (Figure 4.111), *Chrysoperla carnea* is an active predator of many soft-bodied arthropods and their eggs. The genus *Chrysoperla* is

Figure 4.109   Lacewing larva (*Chrysopa perla*) feeding on peach–potato aphid (*Myzus persicae*).

Figure 4.110   Summer form of adult common green lacewing (*Chrysoperla carnea*).

Figure 4.111   Predatory lacewing larva (*Chrysoperla carnea*) attacking rose aphids (*Macrosiphum rosae*).

Figure 4.112   Lacewing egg (*Chrysoperla carnea*) attached to an apple leaf.

an aggregate of closely related species, many of which have been well studied throughout the world. Various species are mass produced in several countries for use on both outdoor and protected crops (e.g., the closely related *C. rufilabris* is often commercially used in North America). Adults are 12 to 20 mm long with a wingspan of up to 30 mm and have long antennae and bright golden eyes. Mature larvae are 10 mm in length.

- *Life cycle:* The optimum conditions for *C. carnea* reproduction are 20°C (68°F) and 80% humidity, with a day length of 15 to 17 hours. Blue/green eggs are laid supported on mucus stalks 4–5 mm long on leaves or plant stems close to their host food (Figure 4.112). These hatch after 3–5 days to produce tiny active predators that feed on honeydew, insect eggs, or small prey. The third

instar larva is extremely voracious and can consume an aphid or a whitefly pupa in less than a minute. The larvae are cannibals and when young may eat unhatched eggs, other larvae, and even adults if food becomes scarce. A silken white cocoon is produced from which the adult emerges. These invariably fly away, contributing little to localized pest control. Overwintering adults turn a pink to brown color (Figure 4.113) and frequently seek refuge in buildings or leaf litter. Special boxes are available to encourage adults to remain within a local area, although a bundle of short bamboo canes with open ends works equally well (see Figure 1.5 in Chapter 1).

- *Crop/pest associations:* In the presence of mixed prey, green lacewings attack aphids first (Figure 4.114), followed by

Figure 4.113   Adult common green lacewing (*Chrysoperla carnea*) showing diapause coloration change from green to brown.

Figure 4.114   Predatory lacewing larvae (*Chrysoperla carnea*) among bird-cherry aphids (*Rhopalosiphum padi*).

thrips and spider mites. They are also known to feed well on young caterpillars and moth eggs, mealybugs, scale insects, whitefly larvae, and pupae. Plants with dense foliage are best suited to these predators, particularly when there is an even spread of prey through the canopy. Frequently in these situations, parasitoids are much less effective and chrysopids have a niche environment.

- *Influence of growing practices:* Due to mass-production systems, lacewing larvae are becoming more widely used to control different pests on a range of crops, including field, garden, orchard, and protected and interior atriums. They are useful on organic crops where pesticide restrictions necessitate a more generalist predator to control many pest species. *C. carnea* are more tolerant to low humidities than other lacewing species.

ORDER: COLEOPTERA (THE BEETLES)

Beetles are recognizable by the hard or leathery forewings (elytra) that often completely cover the more delicate and membranous hindwings. Several species do not possess any hindwings and are thus incapable of flight. However, in all instances, there is a fine line along the dorsal surface marking where the two wing cases join. This is in contrast to the heteropteran bugs, where the leathery wings are folded to overlap, one over the other, often displaying a membranous wing tip. All beetles have biting mouthparts and are able to penetrate the outer skin of most insects and mites. Beetles are typically active hunters with both adults and larvae showing rapid movement. Larvae similarly have biting jaws very much like the adult and frequently eat the same type of food. The larvae of most species have six legs situated in pairs on the first three thoracic segments but, unlike caterpillars, beetle larvae have no prolegs on the abdomen. They may pupate in close proximity to their main food source. Some species produce a simple pupal case on the ground and, in most instances, they are most vulnerable to predation by other organisms while in the pupal stage.

Family: Coccinelidae (Ladybird Beetles)

Adult ladybird beetles range from 0.8 to 7 mm in length and have an oval body with a convex dorsal surface, a flat ventral surface, and biting mouth parts (Figure 4.115). The larva can reach up to 18 mm in length. Coccinellid adults have fully developed hindwings and many species are good fliers. There may be considerable color variation within some species and there are several distinctly recognized forms of many of the more common species. When disturbed, many species simply fall through vegetation and disappear from view in leaf litter. Their colorful wings are a warning signal that they are distasteful to predators such as birds. All species readily produce an acrid tasting, yellow fluid when handled.

Figure 4.115   Adult seven-spot ladybird (*Coccinella septumpunctata*) feeding on black-bean aphids (*Aphis fabae*).

Most commonly known coccinellids are predators of aphids (Figures 4.116 and 4.117). However, other species attack scale insects, mites, mealybugs, and whitefly. Some coccinellids live on fungal spores (Figure 4.118) and are known as fungivores. Coccinellids can be encouraged into agricultural crops by provision of sources of suitable prey in the margins or beneath trees at a time when the crop pest is not present. For example, while many aphids are plant specific and will not attack a neighboring crop plant, many of the coccinellid species will attack aphids on plants in the margin, moving into the cropped area if aphids are present. Proximity of suitable hibernation sites such as grassy banks, hedgerows, or woodland (Figure 4.119) can have a profound influence on the abundance of coccinellids in agricultural areas.

All coccinellids overwinter as adults and usually leave marker scents so that the next generation will be able to use the most successful locations the following year. Many species have only a single generation a year but, under carefully controlled laboratory conditions, can have multiple generations and be commercially mass reared. Most adults only start producing eggs in

Figure 4.116   Young larva of seven-spot ladybird (*Coccinella septumpunctata*) feeding on potato aphids (*Macrosiphum euphorbiae*).

Figure 4.118   Adult 22-spot ladybird (*Psyllobora vigintiduopunctata*) feeding on powdery mildew fungus.

Figure 4.117   *Myzia oblongoguttata*—the striped ladybird—feeding on black-bean aphids (*Aphis fabae*).

Figure 4.119   Adult *Coccinella magnifica*, the scarce seven-spot ladybird, usually associated with wood ants.

midsummer and are attracted to high pest populations, possibly by host sex pheromones. Egg production stops as prey numbers fall and their then lighter weight bodies enable them to disperse to new areas. They are therefore not that useful for release in glasshouses for the control of aphids.

There are few regular commercial producers of ladybirds, the exception being *Adalia bipunctata* for aphid control and those used for control of mealybug and scale insect pests in glasshouses, conservatories, and botanic gardens.

Figure 4.121   Another color variant of *Adalia bipunctata*.

ADALIA BIPUNCTATA (TWO-SPOT LADYBIRD)
The legs and underside of the abdomen of *Adalia bipunctata* are black; markings on the elytra (forewings) are variable, but usually a single black mark is present centrally on each elytron. However, elytra can be almost black with just a few red marks (Figure 4.120). *A. bipunctata* has at least 12 recognized forms, with varying amounts of red and black (Figure 4.121). As one of the most familiar ladybirds, they are 4–5 mm in length, and often found feeding on aphids on trees (Figure 4.122) and shrubs but also on the black-bean aphid (*Aphis fabae*) on broad beans. *A. bipunctata* occurs in field beans but is not common in other arable crops such as cereals. It is diurnal and predominantly active on plants where a source

Figure 4.122   Variants of *Adalia bipunctata* feeding on rosy apple aphid (*Dysaphis plantaginea*).

of food is also present although adults and larvae are found on the ground in late summer when aphid populations are sparse.

This species is commercially available in many countries.

Figure 4.120   Different color variants of the two-spotted ladybird (*Adalia bipunctata*) feeding on rose aphids (*Macrosiphum rosae*).

- *Life cycle:* A. bipunctata overwinters in groups that may be large. Such groups have been found under bark, in buildings and even within houses; there is one record of them being discovered in a larder among neglected cereal packets. In spring, adults emerge from hibernation and find their way out of their overwintering sites. The adults mate in spring and lay their yellow eggs in batches of 5–50 on the lower surface of leaves where they are stacked end on (batches on brassica plants are similar to those of the large white butterfly *Pieris brassicae*). Larvae are slate-gray with orange markings and, when nearly fully grown, are more voracious than the adults. Cannibalism occurs if aphid populations "crash" before the ladybirds have completed development. They hold onto the leaf with an anal sucker, especially when feeding, molting, and pupating. Pupae are black and are found on leaves and stems.

- *Crop/pest associations:* A. bipunctata is common as an aphid predator of black aphids on garden beans and on elder (*Sambucus nigra*), although the aphid on this shrub is toxic to the ladybird. Black-bean aphid (*Aphis fabae*) is also a suboptimal prey and, although the ladybird may not dramatically reduce bean aphid populations, it can accelerate declines, leading to fewer aphids to lay overwintering eggs. On trees, it is a common predator of aphids on lime (*Tilia* spp.) and sycamore (*Acer* spp.).

- *Influence of growing practices:* A. bipunctata benefits from pollen sources in the spring and from other field-margin resources, including aphids on plants such as nettles (*Urtica dioica*). It has a low aphid density threshold for egg laying but may not appear in fields early enough or in high enough numbers to reduce black-bean aphid populations. The slight toxicity of *A. fabae* may be another factor, and proximity of overwintering sites may also affect the numbers entering fields and gardens in the spring.

CHILOCORUS SPP.

Adults and larvae of *Chilocorus* spp. are predators of scale insects in fruit and citrus crops in Europe. *C. nigritus* is an Asian species that provides good control of Diaspid scale insects in tropical glasshouses and conservatories (Figures 4.123 and 4.124); it is commercially available in many countries and licensed for use in the UK. *C. hexacyclus, C. stigma,* and *C. tricyphus* are found in a similar role in North America.

Figure 4.123   Adult *Chilocorus nigritus* feeding on soft scale insect.

Figure 4.124   Adult *Chilocorus nigritus* feeding on soft scale insect.

Figure 4.125   Adult *Chilocorus bipustulatus* on apple leaf.

Adult *Chilocorus bipustulus* are small (3.3–4.5 mm in length) with a very shiny, convex black body and characteristic red spots looking like an exclamation mark in the center of each wing case (Figure 4.125).

- *Life cycle:* Eggs are oval, 1–2 mm long, yellow to orange in color and are laid singly in association with scale insect prey. Larvae and pupae (Figures 4.126 and 4.127) are covered in spines and tubercles and increase in size from 1 to 7 mm in length. *Chilocorus* spp. normally have three to four generations per year in Mediterranean regions and two generations further north. In late summer, typically early autumn in southern France, adult *Chilocorus* spp. leave the orchards and fly to overwintering sites.

- *Crop/pest associations: Chilocorus* spp. are important predators of scale insects on apples, primarily *Lepidosaphes ulmi*, *Epidiaspis leperii*, and *Quadraspidiotus perniciosus*. Individuals typically consume 20–40 scale insects per day.
- *Influence of growing practices:* Like most Coccinellidae, the eggs, larvae, and adults of *Chilocorus* spp. are very susceptible to insecticides. This may be enhanced by the fact that they consume their prey rather than sucking out the body contents, as does a mirid or anthocorid bug. Unfortunately, their scale insect prey is protected by waxy secretions and is much less susceptible to pesticide sprays.

COCCINELLA SEPTEMPUNCTATA (SEVEN-SPOT LADYBIRD)

Coccinella septempunctata adults are 5.5 mm in length or larger (Figures 4.128 and 4.129) and nearly always have three black spots on each red elytron, with one central spot behind the thorax. It is usually found on low vegetation in gardens and arable fields but rarely on shrubs and trees.

Figure 4.126   Larval stage of *Chilocorus* spp. on apple leaf.

Figure 4.127   Pupal stage of *Chilocorus* spp. on apple leaf.

Figure 4.128   Adult seven-spot ladybird (*Coccinella septumpunctata*) feeding on apple aphids (*Aphis pomi*).

Figure 4.129   Adult seven-spot ladybird (*Coccinella septumpunctata*) feeding on mealy cabbage aphid (*Brevicoryne brassicae*).

ORDER: COLEOPTERA (THE BEETLES)

Pale elytra signify a newly emerged adult
(Figure 4.130).

- *Life cycle: C. septempunctata*
  overwinters singly or in small groups,
  in curled leaves of plants such as buxus,
  dry vegetation in gardens and hedge
  bases, and occasionally in clusters on
  the lower branches and stems of low-
  growing shrubs. There is usually one
  generation per year in arable land and it
  is the most frequent ladybird species in
  wheat. Eggs are elongated and smooth
  and laid in small groups (Figure 4.131);
  initially yellow/orange in color, they
  change to light gray prior to larval
  emergence. Neonate (first stage) larvae
  feed on their egg cases (Figure 4.132)
  before dispersing to find aphids on which
  to feed (Figure 4.133). Larval densities
  in midsummer may exceed 100/m^2.
  Larvae are commonly found on the
  soil at this time, moving rapidly on hot
  days to gather as much food as possible,

**Figure 4.131   Eggs of seven-spot ladybird
(*Coccinella septumpunctata*) on cereal leaf.**

**Figure 4.132   Neonate larvae of *Coccinella sep-
tumpunctata* feeding on their empty egg cases.**

**Figure 4.130   Newly emerged adult seven-spot
ladybird (*Coccinella septumpunctata*) next to
empty pupal skin on sweet corn; its color and
spots will develop within a few hours.**

**Figure 4.133   Seven-spot ladybird larva
(*Coccinella septumpunctata*) feeding on winged
(alate) black-bean aphid (*Aphis fabae*).**

Figure 4.134   Seven-spot ladybird (*Coccinella septumpunctata*) pupa on sweet corn.

as aphid populations disappear due to crop ripening. Pupation is often on leaves or stems of plants (Figure 4.134). *C. septempunctata* is often attacked by the parasitoid wasp *Perilitus coccinellae*, and at the end of winter 60%–70% of dispersing adults may be parasitized. The parasitoid's cocoon is spun between the legs of the adult ladybird, which eventually dies, though not immediately, following the parasitoid's emergence. The beetle is paralyzed and dies of starvation or fungal attack, its fat reserves having been consumed by the parasitoid larva.

- *Crop/pest associations:*
  C. septempunctata is common in gardens and especially in cereal fields if aphids are abundant. Late summer swarms, which sometimes are numerous enough to be reported on TV and in newspapers, probably come from ripening wheat fields following outbreaks of the grain aphid *Sitobion avenae*.
- *Influence of growing practices:*
  Insecticide use on wheat in summer

probably kills millions of these insects. If larvae are in the crop, they are either killed by the spray or starve to death when their aphid prey is killed. Herbicides used to "clean up" hedge bases destroy perennial grasses, leaving annual vegetation with a shallower layer of dead and dying vegetation to provide refuges in winter. For the same reason, overly tidy gardens also decrease this species' chances of surviving the winter.

CRYPTOLAEMUS MONTROUZIERI

The mealybug predator *Cryptolaemus montrouzieri* originated in Australia and reached Europe and the rest of the world via California. It controls a wide range of mealybugs. The adult ladybird-shaped beetle belongs to the family Coccinellidae. It is about 4 mm in length, has dark brown/black wing covers, and orange head, prothorax, wing tips, and abdomen (Figures 4.135 and 4.136). Larvae are covered in mealy, wax-like projections, making them (particularly the young larvae) resemble their host mealybug, and they can reach up to 13 mm in length (Figure 4.137). Both the adults and larvae are predatory, with adults and young larvae preferring host eggs and young nymphs, while older larvae will consume all stages of mealybug.

- *Life cycle:* Females mate soon after emergence and start laying eggs about 5 days later; between 200 and 700 eggs

Figure 4.135   Adult Australian ladybird (*Cryptolaemus montrouzieri*).

Figure 4.136   Adult Australian ladybird (*Cryptolaemus montrouzieri*) feeding on young glasshouse mealybug (*Pseudococcus viburni*).

Figure 4.137   Larva of Australian ladybird (*Cryptolaemus montrouzieri*).

are laid singly in mealybug egg masses, so this is not a good species for control of viviparous species such as long-tailed mealybug (*Pseudococcus longispinus*). The number of eggs produced depends strongly on the adult females' diet, and starvation can halt egg production. Several hundred mealybugs may be eaten during development of the predator. Temperature can have a marked effect on the development of *C. montrouzieri*, which takes about 25 days at 30°C (86°F) and up to 72 days at 18°C (64°F); there is no activity below 9°C (48°F).

- *Crop/pest associations:* Larvae move freely over plants in search of prey and, in periods of shortage, can become cannibalistic. Adults under similar conditions fly off in search of more prey. *C. montrouzieri* are also polyphagous,

feeding on aphids and scale insects when necessary. They are commercially produced for sale in most countries and, although they can overwinter in glasshouses, fresh insects are usually introduced for control each spring.

- *Influence of growing practices:* High light levels and bright sunshine play an important part by making the insect much more active. Adults released in an atrium or greenhouse invariably fly directly upward toward the brightest source of light. To counter this behavior, adults should be released at dusk or beneath a fleece or net canopy to reduce their ability to fly away from where they are required to be. The canopy can be removed the following day. The eggs of *C. montrouzieri* are well protected from pesticides in the mealybug egg masses. The larvae and adults are reasonably tolerant of most short-persistence pesticides, providing they are not directly hit.

DELPHASTUS CATALINAE

Adult *Delphastus catalinae* are small, shiny black beetles, 1.3–1.4 mm in length, and are related to the aphid-feeding ladybirds (Coccinellidae). However, this predator devours large numbers of whitefly (Figures 4.138 and 4.139), with both adults and larvae consuming all stages of their host. The pale, yellowish-white larvae are relatively immobile and can starve before

Figure 4.138   Adult *Delphastus catalinae* feeding on glasshouse whitefly (*Trialeurodes vaporariorum*) larval "scales."

Figure 4.139   Adult *D. catalinae* preying on glasshouse whitefly (*Trialeurodes vaporariorum*) larval "scales."

they reach pupation if their food source runs out. *D. catalinae* is distributed widely across the central and southern United States and south through Central and South America as far as Peru. Their use in the UK and parts of Europe requires ministerial license, which is usually granted on the basis of nonsurvival outside the protected environment of a glasshouse.

This species is commercially available in many countries.

- *Life cycle:* Adults require a relatively high pest density of 50–100 whitefly eggs or larvae per square centimeter before oviposition can commence. The adult beetle needs to consume about 150 whitefly eggs or young larvae a day to produce her eggs. Egg-to-adult development is 21 days at 28°C (82°F), which is approximately 3 days slower than the host at similar temperatures. Adult and larvae feed predominantly on young whitefly nymphs. The average longevity is 60–65 days. Diet is important in longevity, as adults can live for up to 80 days feeding on pest larvae, which is significantly longer than males or females feeding on eggs alone.
- *Crop/pest associations:* There is much interest in this predator for suppression of potentially out of control or rapidly breeding whitefly populations,

particularly *Bemisia tabaci* and *Trialeurodes vaporariorum,* on protected crops. *D. catalinae* is commercially used for control of whitefly, especially in localized "hot spots" that frequently occur in amateur glasshouses, where pest numbers can increase to levels uncontrollable by the normal parasitoids. This often happens when wide ranges of individual plants are kept and initial pest levels are neglected; this can result in these small populations building up, unnoticed, to damaging levels. As a first line of defense, routine introductions of parasitoids are suggested with the use of *D. catalinae* as a backup control.

- *Influence of growing practices:* Use of *D. catalinae* in combination with whitefly parasitoids has shown that whitefly containing parasitoid eggs and immature larvae are at risk of predation, but as the parasitoid larvae develop and pupate, the risk is greatly reduced. This is probably due to a hardening of the whitefly cuticle when the parasitoid pupates, making it unpalatable for the predator. The term "biological pesticide" has been coined for this voracious predator, which invariably eats a great deal of prey, but rarely establishes itself in the glasshouse.

HARMONIA AXYRIDIS (HARLEQUIN LADYBIRD)

Native to temperate and subtropical parts of Asia and known by several common names, including "multicolored Asian lady beetle," "Halloween beetle,"' and "Harlequin ladybird," *Harmonia axyridis* can be very difficult to distinguish truly from several other ladybirds. Adults are between 5 and 8 mm in length and extremely variable in color; they are found with light yellow, orange, orange-red, red, or black elytra and spotted with orange-red or black spots. Males tend to have fewer spots (some morphs have none at all). The most common morphs are orange with 15–21 black spots (Figure 4.140) or

Figure 4.140   Orange form of adult harlequin ladybird (*Harmonia axyridis*) next to empty pupal case.

Figure 4.141   Two color variations of adult harlequin ladybird (*Harmonia axyridis*) feeding on peach–potato aphid (*Myzus persicae*).

Figure 4.142   Black color variation of adult harlequin ladybird (*Harmonia axyridis*).

black with two or four orange or red spots (Figures 4.141 and 4.142). The pronotum is straw-yellow to white and has up to five black spots, usually joined to form two curved lines resembling the letter "M."

- *Life cycle:* Eggs are oval, yellow, approximately 1 mm long, and typically laid in clusters of up to 20 on the undersides of leaves. Larvae are often black with an orange band along each side and covered with tiny, soft spines. They have four instars and can reach up to 12 mm in length over a 2- to 3-week period, becoming more voracious feeders as they grow larger. The pupal stage lasts about a week and may be found on leaves, branches, and, occasionally, the sides of buildings before an adult beetle emerges. Adults can live from several months to over a year and overwinter in cracks and crevices of trees and inside buildings.

- *Crop/pest associations:* Although aphids are the main source of food, lepidopteran eggs, psyllids, scale insects, spider mites, thrips, etc., can all be eaten. Most estimates of prey consumption suggest that each complete generation (larvae and adult) is able to kill three to four thousand aphids. They have been introduced, principally for aphid control, on a wide range of crops in several regions throughout the world including North America and parts of Europe. From these introductions they have spread to many countries; they could be found throughout North America by the mid-1990s and several European countries including Britain, where they have become established. Many growers praise this ladybird as an exceptionally useful predator that has been regarded as a crop savior on numerous occasions.

- *Influence of growing practices:* *H. axyridis* are now commonly found in many locations, including several where they have never been permitted for use, such as Canada, the UK, and Scandinavia. However, there are now serious concerns of *H. axyridis* attacking a wide range of nonpest species from different insect orders. They can be a nuisance when they fly to buildings in search of overwintering sites, frequently ending up inside, where they crawl about on windows, walls, etc., before massing together and hiding in cupboards and attics/lofts. If disturbed, they can emit a noxious odor and yellowish fluid, which may stain furnishings, as a defense mechanism against would-be predators.

HARMONIA CONGLOBATA

Harmonia conglobata is an aphid predator in fruit crops, particularly in southern Europe. Adults are large (3.4–5 mm in length) and are recognized by their elytra, which are normally colored pale pink with eight black angular spots, although numerous color variants occur (Figure 4.143). Larvae are gray with pink dorsal spots.

- *Life cycle: H. conglobata* is multivoltine and can produce three or four generations a year in a Mediterranean

climate, and two generations per year in a cooler climate. In early autumn adults leave the orchards and fly to aggregate at overwintering sites, often in prominent locations overlooking the orchards.

- *Crop/pest associations:* Adults and larvae are voracious predators of aphids on fruit trees, particularly *Aphis pomi* and *Dysaphis plantaginea* on apples (Figures 4.144 and 4.145), *Myzus cerasi* on cherries, and *Myzus persicae* on peaches.

Figure 4.144   Adult *Harmonia conglobata* with pink spotted color variation feeding on black aphids.

Figure 4.143   Black color variation of adult harlequin ladybird (*Harmonia conglobata*) feeding on colony of black-bean aphids (*Aphis fabae*).

Figure 4.145   *Harmonia conglobata* adults with color variations including unspotted.

H. conglobata develops and breeds equally well when feeding on either *A. pomi* or *D. plantaginea*. More offspring are produced if adults feed on *M. persicae* or *M. cerasi*. Coccinellids alone seem unable to prevent the summer explosion of *A. pomi*, which occurs when some species (e.g., *Adalia bipunctata*) are in summer diapause and reaches a peak when *H. conglobata* becomes less active. It is also possible that in midsummer *A. pomi* becomes less palatable to ladybirds.

- *Influence of growing practices:* Ladybirds that feed predominantly on aphids are without exception sensitive to insecticides and will be killed by most treatments made to control aphids. However, *H. conglobata* may occur in margins and on trees surrounding an orchard. Care to avoid spray drift and treatment of nontarget habitats would reduce the risk to these predators.

HIPPODAMIA CONVERGENS (CONVERGENT LADY BEETLE)

The convergent ladybird *Hippodamia convergens* is a highly voracious predator, capable of eating 50–60 aphids daily. It can also feed on numerous other insect or mite prey. Adults of this ladybird or lady beetle can vary in size up to 7 mm in length and can have up to 13 black spots on red elytra (Figures 4.146 and 4.147). However, they all have a set of white lines that converge behind the head, giving the common name convergent lady beetle.

This species is commercially available in many countries, but a license for release is required in some.

- *Life cycle:* Each female can produce up to 1,000 eggs at a rate of 10–50 per day over a 1- to 3-month period. These are usually deposited in small clusters close to prey and may be found on leaves or stems. The eggs are about 1 mm long, yellow to orange in color, and spindle shaped. They hatch after 2–5 days to

Figure 4.146   Convergent ladybird (*Hippodamia convergens*) feeding on black-bean aphid (*Aphis fabae*).

Figure 4.147   Convergent ladybird (*Hippodamia convergens*) facing up to an ant (*Messor* sp.). The ant is seeking aphid honeydew, while the ladybird is feeding on apple aphids (*Aphis pomi*); ants will protect aphids from attack by parasitoids and predators.

tiny alligator-like larvae, which begin feeding almost immediately on their own egg cases, followed by insect or mite prey. Depending on food availability and temperature, the larvae can grow to between 5 and 7 mm in length over a 12- to 30-day period, traveling up to 12 m in search of prey. When fully developed, the larva attaches itself by the abdomen to a leaf, stem, string, or other firm surface to pupate. This stage may last from 3 to 14 days, depending on temperature, before the adult emerges.

- *Crop/pest associations: H. convergens* is one of the most abundant *Hippodamia* species in North America and is now widespread throughout most of the world. It was successfully introduced to Chile from California in 1903 to control scale insects. It has recently become available in several European countries, but is prohibited in the UK, which does not allow releases of wild harvested organisms. Each generation (larva and adult) of *H. convergens* is capable of killing up to 5,000 aphids. They can also feed on the eggs and adults of many other soft-bodied insects and mites, including adelgids, beetle larvae, psyllids, mealybugs, scale insects, spider mites, thrips, and whitefly. Ladybirds are susceptible to fungal attack and parasitism, which can reduce their numbers later in the season.

- *Influence of growing practices:* Such active predators tend to eliminate a pest population quite rapidly and find themselves with insufficient food to maintain a constant presence, particularly on protected crops. In field or garden situations, they tend to survive better by feeding on a wider diet and having a better distribution. Currently, *Hippodamia* spp. are collected from wild populations by vacuuming overwintering beetles from aggregation sites (under tree bark and also specially placed shelters). Introduced *H. convergens* (and some other ladybird species) have become a problem in North America, where they have competed with indigenous coccinellid species. They also bite humans when starved. These "harvested" insects need to be fed and conditioned to be ready for use; otherwise, they are likely to fly off in a spring migratory flight before any eggs are laid on the crops to be protected.

PROPYLEA QUATTUORDECIMPUNCTATA (14-SPOT LADYBIRD)

Also known as *Propylea 14-punctata*, the adult has yellow elytra with more or less square-shaped black marks—usually, seven

on each elytron but often fused so that it is difficult to differentiate 14 clear spots. The pronotum (main part of the dorsal surface of the thorax) has an irregular, single black mark (Figure 4.148). Legs are orange-yellow and the lower surface is mainly black. The larva has brighter yellow markings (Figure 4.149) than does that of a similar-sized two-spot ladybird.

- *Life cycle:* P. 14-punctata is a typical ladybird of temperate Europe, with overwintering adults giving rise to one summer generation. The winter adults are not found in large numbers, unlike the aggregations of the two-spot ladybird. It is a diurnal, plant-active species and occurs especially on low-growing plants. *P. 14-punctata* is rarely as abundant as two-spot and seven-spot ladybirds.
- *Crop/pest associations:* P. 14-punctata is most easily found in arable crops; it can

Figure 4.148 Adult 14-spot ladybird (*Propylea quattuordecimpunctata*) on cereal leaf.

Figure 4.149 Fourteen-spot ladybird larva (*Propylea quattuordecimpunctata*) on apple leaf.

ORDER: COLEOPTERA (THE BEETLES)

be as abundant as seven-spot ladybirds but is usually second in importance to this species in terms of biological control of aphids (e.g., in cereals). Enormous swarms of ladybirds occur in late summer, following a mild winter and a hot summer (leading to high aphid populations in cereals); they will include this species, but the seven-spot ladybird dominates. Winter wheat is the crop most likely to harbor *P. 14-punctata,* although it does occur on several garden plants.

- *Influence of growing practices:* Although *P. 14-punctata* can be abundant in arable crops and probably contributes usefully to the reduction in aphid populations, little is known of its overwintering needs or, indeed, of where it spends most of the year outside the 6- to 8-week aphid season in cereals. Insecticide use and mismanaged field boundaries are likely to be unfavorable for this predator, but detailed studies on its ecology have not been carried out.

RODOLIA CARDINALIS (VEDALIA BEETLE)

Rodolia cardinalis is a specialist ladybird predator of fluted or cottony cushion scale insect (*Icerya purchasi*). In 1888 and 1889, following an introduction from Australia, where they originated, of just over 500 adults around Los Angeles, California, they have since been introduced to most citrus-growing areas throughout the world. Adults are 2.5 to 4 mm in length, convex, and red with black spots that may converge to form black patches; they are covered in short, fine hairs, giving a dusty appearance (Figures 4.150 and 4.151).

This species is commercially available in many countries, but a license for release is required in some.

- *Life cycle:* Adults lay 150 to 200 eggs singly or in small groups that hatch after a few days at 25°C to 30°C (77°F to 86°F); the whole cycle from egg to adult takes about 4 weeks. Eggs are red, slightly

Figure 4.150   Mating pair of cardinal ladybirds or vedalia beetles (*Rodolia cardinalis*).

Figure 4.151   Cardinal ladybirds or vedalia beetles (*Rodolia cardinalis*) feeding on cottony-cushion scale insects (*Icerya purchasi*).

oblong, and laid on or close to the egg cases of *I. purchasi* scale insects. These hatch to produce reddish-brown larvae that feed on scale insect eggs; as the larvae develop, they attack and kill all stages of their prey. A red and black pupa is formed within the grayish skin of the final (fourth) larval instar, from which an adult emerges.

- *Crop/pest associations:* A prey-specific predator, *R. cardinalis* can only survive on plants attacked by its host prey, which includes acacia, citrus, olive, and several ornamental trees. Vedalia beetles are now established in most citrus-growing regions of the world and they provide excellent control of cottony cushion scale.

- *Influence of growing practices:* The tremendous success of the California releases is considered the beginning of "classical" biological control because they effectively controlled severe outbreaks of *I. purchasi* in citrus groves. Pesticides, often used to control other pests, can severely disrupt *R. cardinalis*—to such an extent that reintroductions are necessary after ensuring that any persistent pesticide treatments have completely broken down. There are a few small-scale commercial rearing units producing these beetles, mainly for use in restocking pesticide-damaged populations in citrus groves, in botanic gardens and conservatories, and also by the interior landscape industry.

SCYMNUS SUBVILLOSUS

Adults of *Scymnus subvillosus* are small (2.0 to 2.5 mm long), mostly black, with elliptical bronze-colored spots at the front of each

Figure 4.152   Adult lady-beetle (*Scymnus subvillosus*). (Courtesy Frank Koehler, Bornheim, Germany.)

Figure 4.153   *Scymnus subvillosus* larva on apple leaf. (Courtsey Len McLeod/FLPA.)

wing case, and with a covering of fine hairs (Figure 4.152). Larvae are covered in tubercles with white waxy secretions (Figure 4.153).

- *Life cycle:* S. *subvillosus* has two to four generations per year depending on climatic conditions.
- *Crop/pest associations: Scymnus* spp. are minor predators of both aphids and spider mites in apple and pear orchards. They can feed on low populations of aphids such as *Aphis pomi* and tend to be found in higher numbers at the time when aphid numbers are falling. One adult *S. subvillosus* consumes approximately eight aphids per day.
- *Influence of growing practices:* Although useful, naturally occurring predators, they are not commercially produced. Adults and larvae are susceptible to insecticides, both directly and through consuming treated prey.

STETHORUS PUNCTILLUM

Adult *Stethorus punctillum* are small (1.0 to 1.5 mm long) black ladybirds, hemispherical in shape with fine yellow hairs covering the elytra (Figure 4.154). Larvae are dark brown to black and up to 2.5 mm in length with setae on the sides of each segment (Figure 4.155). Both adults and larvae are predators of *Panonychus ulmi* (fruit tree red spider mite) in northern and southern Europe. *S. punctillum* can also be found in

Figure 4.154   Adult *Stethorus punctillum* feeding on two-spotted spider mites (*Tetranychus urticae*).

Figure 4.155   Larval stage of *Stethorus punctillum*.

Figure 4.156   Pair of small black ladybirds (*Stethorus punctillum*) feeding on fruit tree spider mites (*Panonychus ulmi*) on apple leaf.

hedgerows and woodland, where it preys upon other species of spider mites.

- *Life cycle: S. punctillum* overwinters as an adult in crevices and under the bark of trees, usually emerging in late spring. Eggs are laid on the undersides of leaves or occasionally on twigs in early summer, in close proximity to spider mites. Eggs are oval and pale cream colored. In northern Europe a second generation lays eggs in late summer. In southern Europe there may be as many as four generations in 1 year. The development from egg to adult takes approximately 24 days. Adults generally hibernate in autumn on dried leaves on the tree or on the soil.
- *Crop/pest associations:* In fruit orchards, *S. punctillum* feeds entirely on mites, preferring to consume adult prey

(Figure 4.156). Adults feed spasmodically and consume an average of 20 mites per day. Larvae consume an average of 24 mites a day. They provide useful levels of control. During autumn the overwintering eggs of spider mites are eaten by adults and larvae of *S. punctillum*.

- *Influence of growing practices:* All stages of *S. punctillum* are susceptible to insecticides. Winter washes with tar oils in the 1950s harmed overwintering adults and were thought to have contributed to the emergence of *P. ulmi* as a pest species. Similarly, synthetic broad-spectrum pesticides such as the neonicotinoids and pyrethroids, which can persist for several weeks after application, have long-term effects on these and many other predatory insects.

Family: Carabidae (Ground Beetles)

Many ground beetles of the family Carabidae are active polyphagous predators as both larvae (Figures 4.157–4.159) and adults in

Figure 4.157   Predatory ground beetle larva on leaf litter.

Figure 4.158   Predatory ground beetle larva on leaf litter.

Figure 4.159   Predatory ground beetle larva on open ground.

Figure 4.160   Adult predatory ground beetle (*Carabus intricatus*) on moss. Although common in most of Europe, it is considered rare in the UK.

Figure 4.161   Adult predatory ground beetle (*Carabus granulatus*) on decaying wood.

Figure 4.162   Adult predatory ground beetle (*Carabus monilis*) on soil.

Figure 4.163   Adult predatory ground beetle (*Carabus nemoralis*).

Figure 4.164   Adult predatory ground beetle (*Blethsia multipunctata*) inhabits lake shores and bogs.

almost all crops and gardens (Figures 4.160–4.165). Carabids can be found under stones and in damp places in the daytime, and many are active predators at night. While the larvae are almost entirely predatory, the adults of most species consume a mixture of plant and animal food. Some genera, such as *Harpalus* and *Amara*, are largely seed feeders and climb plants in search of food. Carabids are entirely terrestrial beetles and the adults can be recognized by their relatively long legs (adapted for running) and by their filiform (threadlike) antennae. The wing cases cover the abdomen in most species. Flight is not common

Figure 4.165   Adult predatory ground beetle (*Nebria complanata*) on sand.

Figure 4.166   Adult predatory ground beetle (*Agonum dorsale*) on open ground.

and only species in a few genera fly readily. Different carabid species are found in different habitat types, with moisture and soil type being strong determining factors. Although they are useful predators, there is no commercial production of carabid ground beetles. However, the presence of refugia (low-growing shelter plants that may also harbor an alternative source of food) will encourage various species to establish from where they can migrate to other areas. Pitfall traps (see p. 16) are used to monitor the presence and numbers of these ground beetles. The following pages should help with the identification of some of the more common species.

AGONUM DORSALE

Agonum dorsale is a medium-sized (6–8 mm), fast-running beetle with a small, narrow pronotum and long, slender, pale brown legs. Elytra are bicolored, with a brown-orange background and a large, blue metallic patch toward the base (Figures 4.166 and 4.167). It is nocturnal and mainly ground based, but occasionally found on plants when searching for prey. Population density is rather variable between years. Adults are winged.

- *Life cycle:* A. dorsale hibernates as an adult in grassy tussocks, under stones, and often in field margins. It may be found in very dense aggregations in sheltered areas in gardens. It is very active in early spring, moving into fields

Figure 4.167   Adult predatory ground beetle (*Agonum dorsale*) searching among leaf litter.

and laying eggs. Larvae are probably subterranean, with the next generation of adults returning to overwintering sites in the early autumn. Peak abundance in fields occurs in early summer.

- *Crop/pest associations:* These predators are voracious consumers of small invertebrate prey, including aphids that have fallen to the ground or are on accessible plant parts. A large proportion of individuals contain aphids in their guts in the early summer, and populations can aggregate in areas of high aphid density.

- *Influence of growing practices:* A. dorsale is protected from normal winter cultivations and flooding by hibernation away from open ground. It may be affected by spring applications of pesticides that harm overwintering adults as they enter the field. However, this is one of the carabid species that seems to inhabit cultivated land, and there is evidence that populations die out when cultivation ceases.

BEMBIDION LAMPROS

Bembidion lampros is a small (3–4 mm), metallic, bronzy-black beetle with slender legs (Figures 4.168 and 4.169). The most important diagnostic character is the rudimentary maxillary palp, which is only visible under magnification. It is part of a large genus and is widely distributed in cultivated habitats as a day-active, ground-based predator. It is rarely found in dry or sandy soils.

- *Life cycle:* Adults overwinter in hedgerows and grassy tussocks, emerging in the early spring to lay eggs in open crop habitats. Larvae are rarely seen and the new generation of adults emerges from midsummer onward, returning to the hedgerow in early autumn. *B. lampros* is known to enter a state of dormancy, called aestivation, in hotter, dryer conditions in the field in midsummer.
- *Crop/pest associations:* Adults are active soil invertebrate feeders, including aphids, carrot fly eggs, and springtails. They are particularly active in vegetable crops and are ranked highly as an early spring predator, active before other species appear.
- *Influence of growing practices:* Populations can be rather variable

Figure 4.169   Adult predatory ground beetle (*Bembidion lampros*) feeding on an aphid.

between years but their presence in field boundaries and spring breeding may ensure rapid exploitation of crops as long as the adults that have overwintered are not affected by pesticides applied in the spring.

CARABUS VIOLACEUS

Carabus violaceus is a large (24–29 mm), black ground beetle with a blue to purple metallic sheen at the margin of the elytra and with a bluish tinge to the edges of the pronotum (Figures 4.170 and 4.171). The elytra have very fine striations compared with other species of the genus, giving them a matte appearance. Other species of this genus (e.g., *C. granulatus* [see previous Figure 4.161], *C. monilis* [see previous Figure 4.162], and *C. nemoralis* [see previous Figure 4.163]) typically have granulations and characteristic patterns on their elytral sculpture. Like most

Figure 4.168   Pair of predatory ground beetles (*Bembidion lampros*) competing for an aphid.

Figure 4.170   Adult violet ground beetle (*Carabus violaceus*) attacking gray field slug (*Deroceras reticulatum*).

Figure 4.171   Adult violet ground beetle (*Carabus violaceus*) feeding on gray field slug (*D. reticulatum*).

others of the genus, this species is nocturnal but may be found in the daytime under stones or in dense vegetation. Since the adults can inflict a painful bite and can release irritant substances from their anal glands, it is best to avoid handling them.

• *Life cycle: C. violaceus* breeds from midsummer to autumn. Eggs are 5 × 0.5 mm and are laid singly in soil. Larval development takes from 60 to 80 days. The beetles overwinter as larvae. Adults emerge and are active from spring to early autumn and are active on the surface.

• *Crop/pest associations: Carabus* is mainly a Palearctic genus with only a few species occurring in North America or in the Oriental regions. *C. violaceus* is found in Britain and Eastern Europe but, with the exception of the Vosges, is not found in France. This beetle is known to consume slugs, snails, earthworms, and insects in both gardens and agricultural fields.

• *Influence of growing practices:* In northern Europe *C. violaceus* is considered to be a forest species, yet is also commonly found in gardens and in agricultural fields, particularly grass and arable crops once there is a canopy to retain humidity. Large carabids prefer moist habitats and are unlikely to be found in large dry fields with few hiding places. Relatively few individuals of this

genus are collected in field samples, so the effects of crop protection products on them are rarely evaluated.

DEMETRIAS ATRICAPILLUS

Demetrias atricapillus has a long narrow pronotum, with the head the same width. The body is yellowish-red and the head is black (Figures 4.172 and 4.173). It is diurnal and plant active, adhesive tarsi permitting adults to climb cereal and grass stems in search of prey. *D. atricapillus* is very common, especially on grasses and nettles, and is very widely distributed in a variety of crop habitats. It is sometimes found in piles of garden refuse. Some *Demetrias* species are associated with water.

• *Life cycle:* Adults overwinter in grassy tussocks, especially *Dactylis* and *Holcus* species on the boundaries of agricultural fields. They become active in the early spring and migrate into the open crop,

Figure 4.172   Adult predatory ground beetle (*Demetrias atricapillus*) feeding on aphid.

Figure 4.173   Adult *Demetrias atricapillus* feeding on aphid.

climbing foliage of the growing plants. They may migrate up to 100 m. The larvae are rarely seen and possibly subterranean; pupation takes place in the soil with the next generation of adults migrating back to overwintering sites in the late summer.

- *Crop/pest associations:* D. atricapillus is active early in the spring as overwintered adults that forage to build up energy reserves for egg laying. It is best known in cereal crops, where this species seems to specialize in consuming aphids. Early season aphid predation may be important in preventing later outbreaks. Species such as this that overwinter in the field boundary are consuming aphids months before predators like ladybirds colonize crops.

- *Influence of growing practices:* Being diurnal and plant active, D. atricapillus is exposed to direct spraying by pesticides. It seems, however, to be among the most tolerant of all invertebrates to some pesticides, especially synthetic pyrethroids. The most important factor influencing its density in fields is the quality and the proximity of overwintering habitats to the field. Some farmers introduce raised, grassy banks to the centers of fields to ensure that predators such as D. atricapillus penetrate the whole crop at an early stage in the spring.

HARPALUS RUFIPES

Harpalus rufipes is a large (10–17 mm), stout-bodied beetle with short legs. Elytra are covered in a dense, yellowish, velvety pubescence (Figure 4.174). Legs may be pale to dark brown but the head and pronotum are black. A very common species, it is associated with relatively dry areas. Nocturnally active, it is known to climb wheat and other crop plants and feed on aphid colonies.

- *Life cycle:* Both larvae and adults may overwinter deep in the soil, with spring and late summer breeding phases the next year. Most commonly, adults from overwintered larvae are active in the late summer and breeding may continue into a second year. Larvae are weed seed feeders. Adults are winged and are one of the few carabids to be seen flying on warm summer days.

- *Crop/pest associations:* H. rufipes is commonly found in cereal fields and may consume aphids on the ground or on plants. Adults may also feed on mollusks (Figure 4.175) and on strawberry seeds, but rarely cause significant damage.

Figure 4.174 Adult ground beetle (*Harpalus rufipes*), also known as strawberry seed beetle, on cereal leaf.

Figure 4.175 Adult ground beetle (*Harpalus rufipes*) feeding on a slug.

- *Influence of growing practices:*
  H. *rufipes* is relatively protected from sprays when hibernating underground and may rapidly colonize habitats by flight. Population recovery after toxic sprays is usually rapid and H. *rufipes* tend to be numerous every year.

LORICERA PILICORNIS

Loricera pilicornis is a relatively small (6–8 mm) black beetle with protruding eyes (Figure 4.176). The first six segments of the antennae are bristled. The elytra each have three round depressions (Figure 4.177). A relatively common, ground-active predator, it is associated with damper field and garden habitats.

- *Life cycle:* Winged adults may overwinter in wooded areas and hedgerows, but enter fields in the spring. Larvae appear in the spring and are surface active; pupation takes place in the soil. The next generation of adults emerges mid- to late summer and migrates from open fields in the late summer and early autumn.

- *Crop/pest associations:* L. *pilicornis* is an active predator of small invertebrates, including springtails and aphids that are walking on the soil surface. Early spring activity of egg-laying adults may contribute to pest suppression before other predators arrive.

- *Influence of growing practices:* Insecticide application in the spring may suppress overwintering adults or early stage larvae, reducing population size in the next generation. This is one of the ground beetle species that survive best in disturbed or cultivated land and are absent in areas where cultivation has ceased.

NEBRIA BREVICOLLIS

Nebria brevicollis is a large (10–13 mm) brown to black beetle, with long, slender legs (Figure 4.178). The antennae and terminal limb segments may be reddish brown (Figure 4.179) and the pronotum is relatively short. It is a very common field and garden predator associated with damper cultivated areas. N. *brevicollis* scuttles over the soil surface in the autumn when disturbed by cultivation (Figure 4.180).

Figure 4.176   **Head of *Loricera pilicornis*.**

Figure 4.177   **Adult predatory ground beetle (*Loricera pilicornis*) on cereal leaf.**

Figure 4.178   **Adult predatory ground beetle (*Nebria brevicollis*) searching among leaf litter.**

Figure 4.179   Head of *Nebria brevicollis* showing segmented antenna and sharp jaws.

Figure 4.180   Adult predatory ground beetle (*Nebria brevicollis*) searching on open ground.

- *Life cycle:* After summer diapause, adults lay eggs in the soil. Larvae are surface active in the autumn and winter and both are well known to gardeners. Very few adults survive to the end of the winter and the larvae constitute the main overwintering population. This species is thought to have originated from woodland and has a strong association with field margins and hedgerows. The adults are winged, but are rarely seen flying.
- *Crop/pest associations: N. brevicollis* is a voracious predator of aphids and other invertebrate prey on the ground, including mollusks. It is one of the rare winter-active predators and is very common in gardens.
- *Influence of growing practices: N. brevicollis* is affected by autumn/winter spray applications before the crop canopy protects the soil from a full dose

of insecticide. Both larvae and adults may be adversely affected and large population reductions are frequent. Partly as a result of this, populations are variable between years. It is also sensitive to some of the chemicals in slug pellets. This is, however, a species that thrives in association with cultivation, and populations go extinct when cultivation ceases.

NOTIOPHILUS BIGUTTATUS

Notiophilus biguttatus is a small (3.5 to 5.5 mm) black to brassy-colored beetle with obvious protruding eyes and a parallel-sided body form (Figure 4.181). The elytra possess two bright metallic, brassy-yellow stripes. It is a rapidly darting, diurnally active predator.

- *Life cycle:* Adult beetles overwinter and breed in the spring. The next generation of adults is active in the spring and early summer. This beetle is very widely distributed in arable crops and gardens and is also associated with woodland, as are many ground beetles. Some adults have wings, although, as is common in the ground beetles, many do not. This species is considered to be relatively mobile and disperse through a large area. Little is known about larvae, which are probably subterranean, and pupation takes place in the soil.
- *Crop/pest associations: N. biguttatus* is a common predator in spring and summer, eating soil invertebrates, especially springtails and aphids on the ground. It is

Figure 4.181   Adult predatory ground beetle (*Notiophilus biguttatus*) feeding on aphid.

an actively hunting predator and has a pronounced darting gait when disturbed.

- *Influence of growing practices:*
  N. biguttatus is affected by pesticide sprays in the spring and summer, especially because it is diurnal and active at the time of spray application. If the plant canopy being sprayed is dense and runoff of the spray is avoided, the ground beneath the plant may be protected sufficiently for the insect to survive. Its dispersal power suggests that it might rapidly recolonize sprayed sites once residues are ineffective.

POECILUS CUPREUS

Poecilus cupreus is common in agricultural fields throughout Europe, where it prefers relatively moist habitats. The larvae and the adults are polyphagous and consume a variety of prey. Adults of *P. cupreus* are typically bright metallic green in color but may also be bronze or coppery colored (Figure 4.182). The first two antennal segments are a yellow or orange color. *P. cupreus* is a pterostichine carabid (see *Pterostichus melanarius*) with a wedge-shaped body well adapted to pushing and burrowing its way through soil and leaf litter. Adults are 9 to 13 mm in length.

- *Life cycle:* P. cupreus is a spring breeding carabid species that overwinters as an adult. Both adults and larvae consume insect prey, although in England the adults will also attack

young plants of the genus *Beta* such as beetroot, chard, and sugar beet.

- *Crop/pest associations:* P. cupreus is a generalist carabid beetle with no associations known to particular prey types. Common in all field crops, this beetle will consume fly eggs and pupae, aphids, and soil mites. *P. cupreus* thrives in tall, dense vegetation and is often found close to water courses.

- *Influence of growing practices:* Because of its widespread distribution, this species has been used as an indicator for evaluating the effects of pesticides on nontarget arthropods. Adults have been found to be susceptible only to the most toxic insecticides and they are generally considered to be robust predators. Hedgerows and uncropped areas around crops provide an overwintering habitat for many predatory carabid beetle species, including *P. cupreus*.

PTEROSTICHUS MELANARIUS

Pterostichus melanarius is a large (12 to 18 mm), stout beetle with a pronotum that is narrower than the elytra and has long sharp mandibles. It is completely black and generally wingless, with two small punctures on the elytra, and the bead around the pronotum is widened laterally (Figure 4.183). A nocturnal species, it shelters under rocks and vegetation in the day and is completely ground based. *P. melanarius* is

Figure 4.182 **Adult predatory ground beetle (*Poecilus cupreus*).**

Figure 4.183 **Adult predatory ground beetle (*Pterostichus melanarius*) on sand bed.**

very common, found in all kinds of open, damp countryside and in agricultural fields and gardens. A similar predatory carabid, *P. madidus* (syn. *Fabricius madidus*), can be determined by the rounded hind angle to the pronotum compared with *P. melanarius*, which has a sharp hind angle (Figure 4.184).

- *Life cycle:* Eggs are laid in soil and the larvae are subterranean. Adults and larvae overwinter in fields or grassland and migrate in dry periods by walking from open crop environments to damper, grassier habitats. Some adults may breed in a second year. They can cross more wooded areas, but rarely colonize them. Adults are most common in late summer and are particularly abundant in large agricultural fields where egg laying in open field soil may confer an early competitive advantage to larvae.
- *Crop/pest associations: P. melanarius* is an active, voracious predator eating any carrion or invertebrate prey it can locate, including small earthworms (which may dominate its diet), mollusks (Figure 4.185), caterpillars, and aphids on the ground. It may eat strawberry fruit on occasion but is not a serious pest.
- *Influence of growing practices:* Even if the adults are affected by pesticides, the subterranean larvae are protected from most sprays and other harmful practices.

Figure 4.184   Adult *Pterostichus melanarius* (all black, left) competing with *Pterostichus madidus* (red legs, right) for slug.

Figure 4.185   Adult *Pterostichus madidus* feeding on slug.

The most damaging chemicals are in slug pellets, and overuse of these should be avoided in the garden. Because they are wingless, rates of population recovery may be slow.

TRECHUS QUADRISTRIATUS
Trechus quadristriatus is a small (3.0 to 4.0 mm in length), active beetle, brown in color (Figure 4.186) and distinguished from the similarly sized *Bembidion* spp. by the fully developed last segment of the maxillary palps and by the recurrent first elytral stria. It is found on dry terrain with sparse vegetation and also on the moist soil of cultivated fields. *T. quadristriatus* is not confined to agricultural habitats and has been found in mountains, sand dunes, and the nests of small mammals. It is often found flying at night in summer.

- *Life cycle: T. quadristriatus* is an autumn breeder with larval hibernation in the soil, although a small number of adults also overwinter.

Figure 4.186   Adult predatory ground beetle (*Trechus* spp.) on open ground.

- *Crop/pest associations: T. quadristriatus* is known primarily as a polyphagous predator capable of feeding on aphids in cereal crops and is present in fields all the year round. *T. quadristriatus* may be particularly important as a predator early in the season, when many of the colonist species have not yet arrived and when the first cereal aphids are detected.
- *Influence of growing practices:* Although individual *T. quadristriatus* are sensitive to broad-spectrum insecticides such as synthetic pyrethroids, *T. quadristriatus* has been shown to survive in crops treated with insecticides in summer as long as the canopy is fully developed. Since larvae overwinter in the soil, they are generally protected from winter insecticide use.

Family: Cicindelidae (Tiger Beetles)

CICINDELA CAMPESTRIS (FIELD TIGER BEETLE)

Tiger beetles are closely related to ground beetles but have their antennae inserted on the top of the head, just in front of the eyes, and have no striations on their elytra (Figure 4.187). The most common, *Cicindela campestris*, is 10.5 to 14.5 mm in length and bright metallic green with a pale yellow spot on each elytron (Figure 4.188); it can be seen running on the soil surface on sunny days. An adult *C. campestris* will fly readily when disturbed but usually lands after a short distance.

Figure 4.187   Head of *Cicindela campestris* showing pincer jaws and antennae on top of head in front of eyes.

Figure 4.188   Adult green tiger beetle (*Cicindela campestris*) on moss-covered ground.

- *Life cycle:* Adult tiger beetles can be found in fields in warm places from spring to late summer.
- *Crop/pest associations:* Both adults and larvae of *C. campestris* are carnivorous. Larvae make burrows and lie in wait for their prey. They are not commercially produced.
- *Influence of growing practices:* Because of their burrowing habit, tiger beetles are generalist predators found only on light, well-drained soils.

Family: Staphylinidae (Rove Beetles)

Adult staphylinid (rove) beetles are easily identified by their short elytra (forewings), which leave part of their abdomen (usually six segments) exposed (Figure 4.189). Staphylinids can be active polyphagous predators in agricultural crops but some species are fungal and detritus feeders while others are parasitoids of fly pupae (Figure 4.190). The common predatory staphylinids include the larger species, such as the devil's coach horse, *Staphylinus olens* (Figure 4.191), reaching 32 mm in length and found in the daytime under stones in gardens and woodland. Fungal feeders, detritivores, and parasitoid staphylinids are smaller and may be only a few millimeters in length. The smaller species can be extremely difficult to identify and as a consequence they have often been overlooked by entomologists studying their potential as pest control agents.

Figure 4.190   Adult rove beetle (*Paederus littoralis*) on leaf litter.

Figure 4.191   Head of devil's coach horse rove beetle (*Staphylinus olens*) showing jaws.

ALEOCHARA SPP.

Aleochara are small dark brown or black staphylinid beetles common in detritus and moist organic matter (Figure 4.192). Adult beetles fly readily and can be distinguished from other staphylinid families because their antennae are inserted on the upper surface of the head near their eyes. There are a great many species of *Aleochara* and field-collected specimens are difficult to identify. Some can only be identified by examination of their genitalia. The most frequently collected *Aleochara* in arable crops in northern Europe include *Aloconota gregaria*, *Atheta fungi*, and *Amisha* spp. One species, *Aleochara bilineata*, has been mass reared for use in biological control of root maggots. *A. bilineata* has also been used as an indicator species for testing the effects of pesticides and has been widely cultured. It is 5–6 mm in length.

Figure 4.189   Adult rove or staphylinid beetle (*Ontholestes murinus*).

Figure 4.192   Adult rove beetle (*Aleochara bilineata*) on sand bed.

Figure 4.193   Adult *Aleochara bilineata* exiting cabbage root fly (*Delia radicum*) pupal case from which it has just fed.

- *Life cycle: Aleochara* are predatory as adults but many have a parasitoid larval stage. *A. bilineata* lays eggs in moist soil close to plants infested with root maggots. Larvae hatch after 5 days and the first instar larvae enter a fly pupa and begin feeding. *Aleochara* spp. pupate within the host pupa and emerge as an adult after 30–40 days. The time for development from egg to adult takes approximately 6 weeks and there may be two generations per year. *A. bilineata* overwinters as a first instar larva within a host pupa.
- *Crop/pest associations: Aleochara* adults and larvae attack the larvae and pupae of root maggots (Diptera: Anthomyiidae), which are found in a wide range of crops, particularly vegetables. The main host species belong to the genera *Delia* and *Hylemya* and include larvae and pupae of onion fly, wheat bulb fly, and cabbage root fly (Figure 4.193).
- *Influence of growing practices: Aleochara* can typically control 30%– 50% of root fly maggot pupae early in the season and up to 95% late in the season. Broad-spectrum insecticides are harmful to *Aleochara*. Seed dressings give control of the root maggot, while the parasitoid can enter the root to attack the pupa and avoid toxic residues.

ATHETA CORIARIA

The adult *Atheta coriaria* is 3–4 mm long and dark brown to black in color (Figures 4.194 and 4.195). It flies well, but spends most of the time in the soil or growing medium of the crop. Occasionally, it is found on yellow sticky cards, but these seem chance catches. Larvae vary from creamy white in the early instars to a yellowish brown in the later instars (Figure 4.196). All motile life stages are very active and fast moving. They are frequently seen on the soil surface, but readily enter any crevices, often following the plant stem down below the soil line.

This species is commercially available in many countries.

- *Life cycle:* The entire life cycle of the beetle is spent in the soil where females lay their eggs. After hatching, they

Figure 4.194   Adult rove beetle (*Atheta coriaria*).

Figure 4.195   Adult rove beetle (*Atheta coriaria*) with raised tail in defensive mode feeding on blackberry midge larva (*Dasineura plicatrix*).

Figure 4.196   Predatory rove beetle larva (*Atheta coriaria*).

develop through three larval stages, all of which are predatory. The pupal case is composed of soil particles held together by strands of silk. Development time from egg to adult is 21–22 days at 25°C (77°F), decreasing to approximately 11–12 days at 32°C (90°F). Diapause has not been recorded. These beetles can be reared on an artificial diet, which may also have some potential as a supplement to encourage establishment or movement into desired areas in the glasshouse. At high predator populations, cannibalism has been shown to occur.

• *Crop/pest associations: A. coriaria* are generalist predators that feed and complete development on diets of fungus gnats (eggs, larvae, pupae), shore flies (eggs, larvae, pupae), and thrips (pupae,

second-stage larvae). When presented with large numbers of prey, they will kill more than they eat. Other studies have shown them to feed on a wide range of soil-dwelling insects.

• *Influence of growing practices: A. coriaria* have been shown to establish and thrive in glasshouse crops in a number of different growing media, including peat mixes, coconut fiber, and mineral wool. In commercial glasshouses they will readily move into and establish in new plantings. There does not appear to be any effect of irrigation practices (e.g., subirrigation versus top irrigation) on beetle populations. Some insectaries are selling a rearing release system using an artificial diet mixed into a peat and vermiculite media whereby growers purchase a starter colony and produce their own beetles that exit the rearing unit as they mature.

PHILONTHUS COGNATUS

There are 75 species of *Philonthus* in central Europe and they are generally black in color. The largest of the common species in arable crops, *P. cognatus*, is 9.5–10.5 mm in length and has a pale yellow underside to the first antennal segment (Figure 4.197). Other *Philonthus* species are smaller but often have two characteristic lines of three setae on the pronotum with a fourth offset seta to each side. Eyes are smaller than the temple

Figure 4.197   Adult rove beetle (*Philonthus cognatus*).

and the first antennal segment is longer than the distance between the antennae.

- *Life cycle;* Adult and larvae are predatory, attacking a range of arthropod prey. Adults are often found among decaying vegetation from spring to late summer.
- *Crop/pest associations: Philonthus* spp. are among the most voracious of the predatory staphylinid beetles commonly found in cultivated fields and agricultural crops. They have been shown to be one of the major aphid predators in cereal crops and together with other polyphagous predators can limit the size of a cereal aphid outbreak.
- *Influence of growing practices:* Like most of the polyphagous predators, *Philonthus* spp. are adversely affected by the use of broad-spectrum insecticides. Provision of overwintering refugia, such as strips of tussock-forming grasses such as *Holcus lanatus* and *Dactylis glomerata* in and around cereal fields, has been shown to increase the abundance of many predatory beetles greatly.

TACHYPORUS SPP.

Of the *Tachyporus* species found in agricultural crops in northern Europe, *T. hypnorum* is perhaps the most common and the most easily identified, being 4 to 5 mm in length and having a dark circular marking on the pronotum with a pale surround (Figure 4.198). Another common species, *T. chrysomelinus,* has a pale pronotum, as do the less common *T. dispar* and *T. solutus.* *Tachyporus* species are identified primarily by their elytral chaetotaxy (pattern of bristles). Beetles of the genus *Tachyporus* are particularly active in cereal crops at night, where they climb the plant in search of aphids and other prey (Figure 4.199). The characteristic shape of a *Tachyporus* beetle allows it to reach into the space between adjacent cereal grains on the ear and feed on the aphids found there.

Figure 4.198 Adult rove beetle (*Tachyporus hypnorum*).

Figure 4.199 Adult rove beetle (*Tachyporus hypnorum*) feeding on aphids.

- *Life cycle:* Adult *T. hypnorum* can occur in arable crops all the year round but are most numerous in summer. *T. hypnorum* migrates from field margins in mid- to late summer, when breeding takes place. The presence of hedgerows as refugia may be particularly important for this species.
- *Crop/pest associations: Tachyporus* is best known as a cereal aphid predator, particularly active against *Sitobion avenae* in the summer months (Figure 4.200). However, *Tachyporus* species are polyphagous and may also be found in high numbers in orchard crops, gardens, and hedgerows.
- *Influence of growing practices:* Numbers of adults and larvae of *Tachyporus* spp. found in arable crops have declined over the past 20 years. *Tachyporus* spp.

Figure 4.200   Adult rove beetle (*Tachyporus hypnorum*) attacking an aphid.

are also fungivores, feeding mainly on mildews and rusts, so this decline in numbers may be due in part to improved control of foliar diseases as a result of the use of foliar fungicides or the more widespread use of genetic resistance. *Tachyporus* spp. are very dispersive and can recolonize cropped areas each year. Because they are colonists, particularly to arable crops, *Tachyporus* are not generally exposed to autumn insecticides and herbicides.

XANTHOLINUS SPP.

Xantholinus linearis and *X. longiventris* are commonly found in agricultural crops. Both beetles have an elongated body (6 to 9 mm) with a narrow neck and with a rounded shape to the back of the head (Figure 4.201). They are dark in color and have a long first segment to their antennae. Under high

Figure 4.201   Adult predatory rove beetle (*Xantholinus* spp.).

magnification, the thorax of *X. linearis* has a transverse microsculpture (fingerprint-like pattern), whereas *X. longiventris* is smooth.

- *Life cycle: Xantholinus* spp. adults are found in arable crops from early spring and reach a peak in midsummer before declining. Little is known of the life cycles of individual species of *Xantholinus*.
- *Crop/pest associations: Xantholinus* spp. are polyphagous predators in agricultural crops. The adults are relatively slow moving and feed mostly on Diptera larvae and other less mobile prey. Relatively large numbers of Xantholininae can be found in arable crops and grassland.
- *Influence of growing practices: Xantholinus* spp. are susceptible to broad-spectrum insecticides. They can fly and can recolonize crops relatively quickly. Like many polyphagous predators, Xantholininae are not common in very dry habitats. The presence of some ground-covering plants in and around cropped areas will provide refugia with a high humidity, which will help to support predatory species.

ORDER: DIPTERA

Insects of the order Diptera have one pair of membranous wings and a pair of minute, club-shaped balancing organs called halteres. Adults feed on a liquid diet from decaying matter, insect honeydew, plant nectar and pollen, or, in some species, blood. Most (primitive) larvae have biting mouthparts or, in advanced species, a pair of specialized hooks that rupture the surface skin from which they can suck the liquid contents. Larvae lack true legs, although some species have fleshy stumps that aid their wriggling movement, similarly to the prolegs of Lepidoptera.

Many dipterous flies are predatory at some stage in their life cycle (e.g., the larvae of

Figure 4.202 Head view of adult female predatory midge (*Aphidoletes aphidimyza*) showing compound eye and antennae.

Figure 4.203 Adult predatory dance fly (*Empis stercorea*); note small, club-shaped halteres.

Aphidoletes aphidimyza are predatory against many species of aphid while the delicate adult feeds only on honeydew or nectar) (Figure 4.202). Both adult and larva of *Empis tessellate* are predatory, but on different hosts for each life stage. Several of the larger Diptera, such as hoverflies, mimic bees and wasps with a striped thorax; they do not have a constricted waist and only one pair of wings, giving a clear indication of their taxonomy.

Family: Empididae (Dance Flies)

EMPIS STERCOREA AND *EMPIS TESSELLATA*

There are over 3,000 known species of the Empididae family of flies and, as they have a vast range in size and form, they can be easily mistaken for other Diptera (Figure 4.203). Both adult and larvae are predatory on other insects, mainly feeding on smaller, adult flies and larvae; however, they are also frequent visitors to flowers for nectar. Larvae may be found in a wide range of habitats from freshwater ponds to many that live on prey feeding on animal dung, decaying plant matter, and in soil. The common name "dance flies" arises from their courtship displays involving the presentation of prey as a stimulus to mating. At 9 to 11 mm in length, the adults of *Empis tessellata* are among the largest members of the family (Figure 4.204). The wings are strongly yellowish-brown tinted and the adult fly is conspicuous because of its size and dark gray color. The maggot-like

Figure 4.204 Adult predatory dance fly (*Empis tessellata*) among aphid colony.

larvae are creamy yellow and feed with internal mouth hooks; they reach 13 to 14 mm when fully grown.

- *Life cycle:* E. tessellata occurs throughout Europe; other members of the genus *Empis* are found throughout the world. The adult flies occur through the summer months and are common and widely distributed in fields, pastures, wetlands, and lightly wooded areas. The larvae live in the soil and leaf litter and, like the adults, are also predatory, feeding on other soft-bodied arthropods. The diet of the larvae consists most probably of other fly larvae that occur commonly in soils. Little is known of the biology of E. tessellata including the overwintering stage.

- *Crop/pest associations:* Many empid flies, including *E. tessellata*, have been reported as important predators of agricultural pests, principally of flies such as leaf miners and other midges but also of other plant-feeding insects such as aphids and plant bugs. However, their precise role has not been fully documented. They are not commercially produced.

- *Influences of growing practices:* Little is known of the biology of *E. tessellata* or other empids, but cultivated soils will frequently upset their breeding sites. The adult flies are susceptible to pesticide sprays.

Family: Muscidae (Hunter Flies)

COENOSIA ATTENUATA

Coenosia attenuata are small predatory flies that are similar in appearance and from the same family (Diptera: Muscidae) as houseflies (Figure 4.205). Adult males are 2.5–3 mm in length and pale blue-gray in color, whereas females are slightly larger and darker gray to black. The common name "hunter fly" is apt as the adults actively attack flying prey as they pass nearby, catching most small, flying insects such as scatella flies (Figures 4.206 and 4.207), sciarid flies, moth flies, and whiteflies. Originally from southern Europe, they are found in most countries of the world; it is assumed they arrived on imported plant material along with their prey.

- *Life cycle:* One complete generation can take as little as 25 days at 28°C

Figure 4.205   Adult predatory hunter fly (*Coenosia attenuata*) at rest on leaf.

Figure 4.206   Adult predatory hunter fly (*Coenosia attenuata*) feeding on adult fungus fly (*Scatella stagnalis*).

Figure 4.207   Adult predatory hunter fly (*Coenosia attenuata*) feeding on adult fungus fly (*Scatella stagnalis*).

(82°F) to 40+ days at 18°C–20°C (62°F–68°F), but they can tolerate much lower temperatures. Adults can live for an average of 3–4 weeks and lay their eggs in groups of four to six, depositing them within small pockets of compost or soil. Studies indicate that eggs hatch after 5–7 days directly to a third instar larval stage and develop on to a fourth before pupating in the soil. Larvae are soil dwelling, feeding on the larval and pupal stages of most other insects that inhabit the same environment; they can be cannibalistic.

- *Crop/pest associations:* C. attenuata are polyphagous predators of many flying insects and can make a significant contribution in a biological control program. However, they will also attack flying parasitoids as well as pest species. They are most likely to be found in glasshouses where there are pot plants and soil-grown salad crops. Slower growing plants such as begonia, New Guinea impatiens, and poinsettia that are susceptible to attack by scatella and sciarid flies favor this predator. Large populations can establish in most moist environments where suitable prey also exists. Dead prey insects may be found on the upper surface of leaves on which the predator has fed but these do not usually create a problem as they are easily dislodged in transit.

- *Influence of growing practices:* The larval stages can be particularly difficult to find in soil; usually, the first sign of their presence is adult flies on yellow sticky traps. This is because they rest on a flat surface (vertical or horizontal) while they wait for any flying insect to pass close by. If a small piece of tissue paper or cotton lint is dropped a few centimeters in front of the adult hunter fly, it will invariably pounce on it and almost immediately release it once it finds that it is not an edible prey. The soil-living larval stages are likely to be susceptible to parasitic nematodes used to control sciarid larvae by direct mortality and lack of host prey. Adult populations of sciarids and other pest flies can be reduced by the use of long-persistence pesticides such as synthetic pyrethroids, which will inevitably reduce predator numbers.

Family: Tachinidae (Parasitoid Flies)

Tachinid flies are members of the insect order Diptera, although they are extremely diverse in appearance. The majority look like bristly houseflies and are 6 to 20 mm in length (Figure 4.208). Some species are drab gray, with or without light stripes, while others can be brightly colored. They are found in

Figure 4.208   Adult parasitic tachinid fly (Tachinidae) depositing white eggs on caterpillar. (Photograph courtesy of Larry West/FLPA.)

vegetated areas throughout the world, where they provide useful control of many insect pests. Adults feed on plant nectar while the larval stages are parasitic on a range of insects—commonly, Lepidoptera larvae but some species attack adult and larval stages of beetles, grasshoppers, and sawflies. Most tachinids have a broad range of target hosts; however, a few are specific against a single species of host.

- *Life cycle:* Tachinid flies' elongated white eggs can be deposited directly on the host or inserted within the host or in clusters on the host's food that may be ingested or hatch and attack at a later stage. In some species the eggs hatch within the adult fly, giving live birth to larvae. This tactic is mostly found in tachinids that attack plant-stem-feeding insects whereby the parasitoid larvae must rapidly locate its host within the plant to survive. Others deposit quite mature eggs that hatch almost immediately to an active larva that rapidly penetrates the target host before it has a chance of dislodging the egg. Developing parasitoid larvae initially feed on body fluids and nonessential tissues as the host continues to grow normally; there may be visible external signs of attack in the form of small dark lesions visible through the host skin. In the later stages of development, the parasitoid larva consumes the essential organs of the host, eventually killing it. Pupation may occur externally or within the host cadaver with one, two, or more flies coming from each host. From spring to late summer the adults emerge after just 1–2 weeks, having multiple generations each year; however, as autumn proceeds, the larvae overwinter within the host's overwintering stage as either larvae or pupae depending on host species.

- *Crop/pest associations:* While a few species have been introduced as biocontrol agents on agricultural crops and forest plantations, most are naturally occurring in gardens and open spaces with mixed flowering plants. Currently, there are no commercial rearing systems, although laboratory methods have been developed for specific species against specific economic pests. From 1920 to the 1980s, the tachinid *Lydella thompsoni* was introduced to North America to control the European corn borer (*Ostrinia nubilalis*); it is now considered established in several states.
- *Influence of growing practices:* Adult flies require a source of plant nectar, so large areas of monocropped land tend to support fewer flies than areas with mixed early flowering plants. Broad-spectrum, long-persistent pesticides can severely deplete numbers of host insects as well as these parasitic flies, often resulting in further use of pesticide treatments. Although extremely useful parasitoids of many other insect species, they are considered a pest of silkworm (*Bombyx mori*) farms and butterfly houses, where they indiscriminately attack the caterpillar larvae.

Family: Cecidomyiidae (Predatory Midges)

The majority of insects in the family Cecidomyiidae are gall-forming midges capable of considerable plant damage. However, some are notable predators, particularly of aphids and spider mites. Many species are found throughout the world and have been extensively studied for their commercial value as both mass-produced and naturally occurring predators. The larvae are the only predacious stage and can kill far more prey than they consume, which makes them an extremely useful control agent. Adult midges produce more viable eggs when they feed on pest honeydew or plant nectar than when water alone is available.

APHIDOLETES APHIDIMYZA

Adult *Aphidoletes aphidimyza* closely resemble miniature mosquitoes, having long legs and delicate wings. Females have bead-like antennae (Figures 4.209 and 4.210), while the males' antennae are feathery and almost cover the length of the body (Figure 4.211). Adults are nocturnal but can be found hanging in shady places on plants and pots during the day. Larvae are the most noticeable stage, with their orange coloration (Figures 4.212 and 4.213); under magnification white, fat bodies can be clearly seen. *A. aphidimyza* feeding on black aphids have a darker orange appearance due to internal staining from the host's body fluids.

This species is commercially available in many countries.

- *Life cycle:* Oval, shiny orange/red eggs (0.3 × 0.1 mm) are deposited close to aphid colonies. On hatching, the larvae

Figure 4.209   Adult female predatory midge (*Aphidoletes aphidimyza*).

Figure 4.210   Adult female *Aphidoletes aphidmyza* (left) facing winged aphid (*Aphis gossypii*).

Figure 4.213   Predatory midge larva (*Aphidoletes aphidimyza*) feeding on mint aphid (*Ovatus cratae- garius*).

Figure 4.211   Adult male *Aphidoletes aphidimyza* with large, feathery antennae.

Figure 4.214   Empty pupal skins of *Aphidoletes aphidmyza* from which the adults have emerged.

Figure 4.212   Predatory midge larvae (*Aphidoletes aphidimyza*) feeding among a colony of chrysanthemum aphids (*Macrosiphoniella sanborni*).

are of similar size to the egg and for the first few days may be difficult to find among aphid colonies. The young larvae feed on honeydew and also directly on the host aphids. Before feeding on an aphid, it is paralyzed by

an injection of toxin, usually in the leg. The larva then begins sucking out the juices from the body of the aphid. After 7–10 days the larva is about 2.5 mm in length and clearly visible among the aphids. When fully grown, the larva drops from the leaf and forms a silken cocoon that is often covered in grains of sand and within which it pupates. Emergence is usually about 10–14 days later—providing the pupating medium does not desiccate and leave an empty, transparent skin (Figure 4.214).

- *Crop/pest associations: A. aphidimyza* is predatory on all the common glasshouse aphids and many outdoor species. Prey consumption ranges from 5–10 large aphids to up to 80–100 smaller aphids per larva during its life time. Aphids die after being attacked by the predator larvae, with many aphids killed by injection of toxin but not eaten. Dead aphids remain attached to the leaf but frequently fall off when they begin to decay, leaving a relatively clean plant.

- *Influence of growing practices:* The adults are nocturnal and require a period of darkness for mating and oviposition. Consequently, in areas of continuous lighting such as airports, shopping malls, and some office atriums, this predator does not work well. Conversely, the larvae require at least 15.5 hours of light (either natural, i.e., from midspring to late summer, or from an artificial source) to prevent the pupae from entering diapause, although very low-intensity light has been shown to be sufficient for this purpose. Adults are more susceptible to pesticides than larvae and the generally selective aphicide pirimicarb is also harmful to the adult. Banker plants of cereals, such as barley, rye, and wheat, are often infested with the cereal aphids (*Rhopalosiphum padi* or *Sitobion avenae*) and such infested plants produce many beneficials *in situ*.

FELTIELLA ACARISUGA

Feltiella acarisuga is visually similar to the aphid predatory midge *Aphidoletes aphidimyza* in both adult and larval stages (Figures 4.215 and 4.216), and like *A. aphidimyza*, only the larvae of this cecidomyiid midge are predatory (Figure 4.217). However, unlike *A. aphidimyza*, *Feltiella acarisuga* pupates on the leaf in a white silken cocoon (Figure 4.218). It is this habit that has caused restrictions in the mass production for commercial use of this insect due to higher labor costs; however, it is commercially available in many countries.

- *Life cycle:* Orange/red eggs are laid among spider mite colonies that, after 3–5 days, hatch to minute orange-colored larvae. These feed on all stages of spider mite and can eat up to 15 eggs,

five young mites, or three mature mites each day. After a further 5–7 days at 22°C (72°F), the fully developed larva spins a silken cocoon, usually against the side of a leaf vein, in which a pupa is formed. White silken cocoons are

Figure 4.215   Adult predatory midge (*Feltiella acarisuga*).

Figure 4.216   Young *Feltiella acarisuga* larva feeding on two-spotted spider mite (*Tetranychus urticae*).

Figure 4.217   Mature *Feltiella acarisuga* larva feeding on two-spotted spider mite (*Tetranychus urticae*).

Figure 4.218   Silk cocoon of predatory midge (*Feltiella acarisuga*) next to leaf vein.

Figure 4.219   Adult *Feltiella acarisuga* emerging from cocoon.

generally found on the undersides of leaves; an almost transparent skin protruding from one end indicates that the adult has emerged (Figure 4.219).

- *Crop/pest associations:* Little is known about the fecundity or host-searching ability of this predator other than it appears in large numbers some years and rarely in others. In most instances it is found where biological control is being used for other pests and pesticide usage is restricted. *F. acarisuga* can be parasitized by an *Aphanogmus* species of wasp that is likely to be responsible for the large population fluctuations seen between years. The parasitoid wasp attacks the larval stage and emerges as an adult from the cocoon (Figures 4.220 and 4.221). This makes it difficult to distinguish between healthy predator cocoons and those carrying a parasitoid.

- *Influence of growing practices:* *F. acarisuga* prefers humidity of about 85% but is more effective and tolerant to low humidities than *Phytoseiulus persimilis* (see p. 84). The presence of aphid or whitefly honeydew as a food source for adults has resulted in increased fecundity. As the predator pupates within a cocoon attached to a leaf, heavy deleafing and leaf removal of glasshouse tomato crops may deplete their numbers. Combined introductions

Figure 4.220   *Feltiella acarisuga* cocoon showing drill holes made by parasite (*Aphanogmus* sp.).

Figure 4.221   *Feltiella acarisuga* cocoon opened to show pupating *Aphanogmus* sp. parasitoid wasp with spherical frass pellet.

with polyphagous predators such as *Chrysoperla* and *Macrolophus* spp. are not recommended, as the slow-moving larvae frequently fall victim to these predators.

Family: Syrphidae (Hoverflies)

Figure 4.222   Adult hoverfly (*Syrphus ribesii*) on apple blossom.

Predatory hoverflies are closely associated with woodland and flowering habitats (Figures 4.222–4.224) but have adapted to man-made environments, where they are commonly found in gardens, and agricultural crops where aphids occur. Adult hoverflies feed primarily on nectar and pollen but will also consume aphid honeydew. Larvae of several hoverflies are voracious predators of aphids. However, other hoverflies have very diverse larval feeding habits; these include detritus-feeding aquatic rat-tailed maggots that develop into drone flies (such as *Eristalis tenax*) (Figures 4.225 and 4.226), the narcissus bulb fly (*Merodon equestris*) (see Figures 3.75 and 3.76 in Chapter 3), and bumblebee mimics such as *Volucella bombylans* that live inside bee nests, feeding on rubbish (and most likely the occasional bee larva). Only hoverflies of the subfamily Syrphinae are aphid predators, often attacking a wide range of aphid species (Figures 4.227–4.230). Adults of the more common hoverflies are easily identified by the patterns of their body markings.

Most hoverflies are mimics of bees and wasps and share the aposematic (warning) coloration of many of these Hymenoptera. They are, however, true flies (order: Diptera) and are therefore the only insect order to have one pair of wings.

Figure 4.223   Adult hoverfly (*Episyrphus balteatus*) on Compositae flower.

Figure 4.224   Adult hoverfly (*Sphaerophoria scripta*) on yarrow flowers (Umbellifera).

This fly family has a characteristic "false vein" (*vena spuria*) in the wing that can be seen with the aid of a hand lens or a binocular microscope.

Figure 4.225   Adult drone flies (*Eristalis tenax*), taking pollen from a Compositae flower, are good pollinators.

Figure 4.228   Hoverfly egg on reedmace leaf with colony of water lily aphid (*Rhopalosiphum nymphaeae*).

Figure 4.226   Rat-tailed maggot of drone fly (*Eristalis tenax*) living in wet compost.

Figure 4.229   Predatory hoverfly larva feeding on water lily aphids (*Rhopalosiphum nymphaeae*).

Figure 4.227   Predatory hoverfly larva feeding on rose aphids (*Macrosiphum rosae*).

Figure 4.230   Hoverfly pupal case; the adults head develops in the broad end of the case.

EPISYRPHUS BALTEATUS

The abdomen of *Episyrphus balteatus* is banded black and dull orange (Figures 4.231 and 4.232). The eyes of the male touch on the dorsal surface of the head; those of the female are separated. The adults hover over flowers and aphid colonies.

- *Life cycle:* Both adults and the pupal stage overwinter. Both sexes take pollen and nectar from wild and garden plants from early spring onward (an important aspect for their enhancement in crops). The females need the amino acids from pollen to mature their

Figure 4.231   Adult hoverfly (*Episyrphus balteatus*) on Compositae flower.

Figure 4.233   Egg of *Episyrphus balteatus* among black-bean aphids (*Aphis fabae*).

Figure 4.232   Adult hoverfly (*Episyrphus balteatus*) ovipositing among rose aphids (*Macrosiphum rosae*) on rose.

Figure 4.234   Late instar larva of *Episyrphus balteatus* among colony of aphids.

eggs, while both sexes use the nectar for energy. Eggs are laid in or close to aphid colonies (Figure 4.233). Maggot-like larvae hatch in a few days and voraciously consume aphids, several hundred of these being eaten during larval development (Figure 4.234). Larvae are mainly nocturnal feeders, so they are not so easily seen during the daytime, as they shelter behind leaf sheaths. Pear-shaped pupae (Figure 4.235) are fixed to the plant and the adults emerge after a few days. Spring adults may give rise to two more generations before autumn. Most gardeners see hoverfly adults on or near open flowers as these short-tongued insects can easily obtain nectar and pollen from such plants, especially those in the families Umbelliferae, Chenopodiacae, and (pollen only) Poaceae (grasses). They can be very abundant in late summer, and like the seven-spot ladybirds, adults emigrate from ripening cereals that had earlier supported high aphid and predator populations.

Figure 4.235   Pupal case of *Episyrphus balteatus*.

- *Crop/pest associations:* Some hoverflies feed as larvae on decaying organic matter or occur in stagnant water, but the species useful in biocontrol (including *E. balteatus*) are aphid feeders as larvae. *E. balteatus* larvae are common in arable crops, as well as in horticulture and gardens, and feed on a wide range of aphid species. They are less common on shrubs and trees, although hoverflies do occupy this niche.
- *Influence of growing practices:* *E. balteatus* will lay eggs at very low aphid densities, but "insurance" pesticide sprays will frequently kill them. Pyrethroid insecticides have a powerful insect deterrent property that inhibits flying adult insects from sprayed areas. Because of this they are relatively less harmful than some organophosphate and carbamate products. The dependence of the adults on floral resources means that unmanaged field boundaries in which annual weeds take the place of other dicotyledonous plants offer fewer resources in spring. Drilling field margins with strips of pollen-rich plants, such as tansey-leaf (*Phacelia tanacetifolia*) or buckwheat (*Fagopyrum esculentum*), or Umbelliferae such as coriander (*Coriandrum sativum*) can enhance populations of adult hoverflies in adjacent fields, resulting in a higher rate of egg laying and of aphid predation.

SCAEVA PYRASTRI

Adult *Scaeva* species are large and distinctive with three pairs of white markings on a black abdomen, the middle pair crescent shaped (Figures 4.236 and 4.237). The larvae are aphid predators and feed on a wide range of prey species. Eggs are oblong, white in color, and about 1 mm in length.

- *Life cycle:* *Scaeva pyrastri* is multivoltine, with three to four generations per year and with the adult females overwintering. In mild seasons it is not uncommon to see these flies active throughout the winter months. Eggs are laid among aphid colonies from early spring onward. Larvae develop as they feed on the aphid colony and pupate on the ground beneath the crop. Emerging adults often migrate and can cover long distances in search of suitable aphid-rich sites for egg laying.
- *Crop/pest associations:* Because it overwinters as an adult, *Scaeva* spp. can appear early in the season and lay eggs among the first aphid colonies to appear in crops. Although larvae attack many aphid species, *Scaeva* spp. are considered to be important predators

Figure 4.236   Adult hoverfly (*Scaeva pyrastri*) on peppermint flower.

Figure 4.237   Adult hoverfly (*Scaeva pyrastri*) on Compositae leaf.

Figure 4.238   Adult hoverfly (*Syrphus ribesii*) on apple blossom.

Figure 4.239   Adult hoverfly (*Syrphus ribesii*) on apple leaf.

of *Rhopalosiphum insertum* in fruit orchards early in the season, and of *Dysaphis plantaginea* from spring to midsummer.

- *Influence of growing practices:* As for all hoverflies; the provision of sources of pollen and nectar helps adults to feed and to mature their ovaries and produce eggs. Broad-spectrum insecticides used against aphids are toxic to adults and larvae of *S. pyrastri* and other hoverfly species. Selective aphicides, such as pirimicarb and pymetrozine, are less harmful to larvae and adults of hoverflies.

SYRPHUS RIBESII

Syrphus ribesii is a widespread and common hoverfly species, with characteristic yellow abdominal markings (Figures 4.238 and 4.239). Larvae can be identified by the shape of their breathing tubes and by their coloration, but are best recognized by the

adult stage. Adult males, stationary on twigs or foliage, can produce a high-pitched sound, using their flight muscles to raise their body temperature.

- *Life cycle:* Females lay eggs in or near aphid colonies during the summer. Larvae are voracious predators of aphids in many agricultural crops and in gardens, piercing the aphid's body and sucking the contents dry. While feeding, the larvae often raise their mouthparts and lift the aphid off the plant. The larval stage of *S. ribesii* can be as short as 10 days and under favorable conditions there can be as many as three generations per year. It overwinters mostly as a larva on the ground or hidden in vegetation. In spring the larva pupates and after a few days the adult emerges. However, in autumn many adults of *Syrphus* species avoid winter by migrating south to the Mediterranean.

- *Crop/pest associations:* *S. ribesii* is an important aphid predator in orchards and extensive agricultural crops. It also occurs in gardens. An individual larva can eat more than a thousand aphids during its development. *S. ribesii* adults have particularly short mouthparts, so they can only feed on pollen of very open flowers (Figure 4.240).

Figure 4.240   Adult hoverfly (*Syrphus ribesii*) dusted with pollen from *Phacelia tanacetifolia* flower; these plants are often used to encourage beneficial insects.

- *Influence of growing practices:* Provision of flowering plants as pollen and nectar sources has been shown to encourage hoverflies. Compositae and Umbelliferae are thought to be the most useful plant families for this purpose. Two species with a high availability of pollen and nectar for the flies that are readily available as seeds are *Phacelia tanacetifolia* and *Fagopyrum esculentum*. Hoverfly larvae are sensitive to most broad-spectrum insecticides but are not generally affected by fungicides.

ORDER: HYMENOPTERA (PARASITOID WASPS)

The Hymenoptera is an extremely large order containing many commonly known insects, such as ants, bees, and wasps. They usually have two pairs of membranous wings. The rear pair is smaller than the front pair and has a row of minute hooks along the front edge that attach the two together. Due to reduced venation and large cells, the wings may appear transparent. There are two suborders: the Symphyta and Apocrita. The Symphyta contains wood-boring and leaf-feeding sawflies whose adults have no waist, but the caterpillar-like larvae have well developed heads and more than four prolegs on the abdomen but without the crochet hooks characteristic of Lepidoptera.

Larvae of the Apocrita are apodous (without legs) and have a reduced head capsule. They are generally either parasitic within other insect species or are solitary or social insects (ants, bees, and wasps) where the larvae are provided with all the food they need. Adults have a constricted waist between the thorax and abdomen, giving the characteristic wasp appearance (Figure 4.241). The waist is formed by a ring-like second abdominal segment (petiole) that gives considerable

Figure 4.241 Adult (*Aphidius ervi*) preparing to oviposit into young mottled arum aphid (*Aulacorthum circumflexum*).

Figure 4.242 Adult *Aphidius colemani* preparing to insert an egg (oviposit) into an aphid.

abdominal flexibility (see *Aphidius* ovipositing, Figure 4.242), enabling adult wasps to oviposit in a confined space. The larvae are legless and have a reduced head with relatively simple mouthparts. This is due to the close proximity of their food source as seen in the gall-forming plant feeders that are surrounded by their food and with the ectoparasitoid (*Diglyphus isaea*), where

the larva develops directly next to its host or, most commonly, as an endoparasitoid within the host body (*Aphidius* and *Encarsia* spp.) (Figure 4.243). Mature larvae spin a silken cocoon within the host body to form a "mummy" (named after Egyptian mummies that contain a body wrapped in bandages) in which it pupates (Figure 4.244). Most mummies are golden brown in color, although a few are black (see Figure 4.253 later in this chapter), which may speed up development by better utilization of solar heat.

There are two divisions within the Apocrita: the Aculeata, to which ants, bees, and wasps belong and characterized by having their ovipositor modified as a sting, and the Parasitica, which are nearly all parasitoids

Figure 4.243 Adult Hymenoptera parasitoid wasp (*Encarsia formosa*) among parasitized (black) whitefly pupae (*Trialeurodes vaporariorum*).

Figure 4.244 Swollen body of pea aphid mummy parasitized by *Aphidius ervi* (on left) compared to healthy pea aphid (*Acyrthosiphon pisum*) on right.

with ovipositors adapted for piercing host tissues.

Many parasitoids extend their longevity and fecundity by feeding on host hemolymph after causing a wound with their mandibles or, more commonly, with their ovipositor. Host feeding can result in as many individual hosts killed by adult feeding as by parasitism. This can be viewed as a form of predation. When the ovipositor is used to inflict a wound, the adult parasitoid may feed directly as hemolymph oozes out. Others puncture the host and allow the fluid hemolymph to coagulate around the ovipositor, forming a straw-like tube to feed from. This occurs when the host is hidden within plant tissue. Host feeding is usually necessary for continued egg production and sustaining oviposition. Adult parasitoids that have been starved (host scarcity or, in the case of commercial supplies, those that have been delayed in transit for several days) can resorb their eggs to maintain longevity and must host feed to resume oviposition.

Multiple parasitism occurs when females of more than one species oviposit in the same host individual, as may occur among aphid colonies being attacked by different parasitoids. Similarly, superparasitism occurs when more eggs of the same species are deposited to an individual host than can reach maturity, regardless of whether laid by a single or several females. Polyembryony, on the other hand, is the development of multiple individuals from a single egg that in some species can produce over 1,000 larvae. Hyperparasitism occurs when one parasitoid species develops as a parasitoid of another parasitoid; an example is shown with an *Aphidius colemani* mummy. Compare the smooth exit hole cut at the rear of the host from the primary parasitoid (Figure 4.245) to the jagged edge exit hole at the top of the host produced by the hyperparasitoid *Asaphes* spp. (Figures 4.246 and 4.247).

Figure 4.245 Exit holes cut through *Aphidius colemani* parasitized mummies of cotton aphid (*Aphis gossypii*).

Figure 4.246 Exit hole cut through *Aphidius* spp. mummy by hyperparasitoid wasp (*Asaphes* spp.). (Photograph courtesy of Dr. Rob Jacobson, UK.)

Figure 4.247 Adult female *Asaphes* spp. hyperparasitoid ovipositing into *Aphidius* spp. mummy. (Photograph courtesy of Dr. Rob Jacobson, UK.)

Figure 4.248   Adult parasitoid wasp (*Anagrus atomus*) ovipositing in eggs of glasshouse leafhopper (*Hauptidia maroccana*).

Figure 4.249   Orange/red eggs of *Hauptidia maroccana* containing developing *Anagrus atomus* parasitoid wasps.

Family: Mymaridae

ANAGRUS ATOMUS

Anagrus atomus is a minute orange/red wasp (0.6 mm long) with feathery wings (also 0.6 mm long when fully extended) and long antennae (Figure 4.248). The adults are very delicate and short lived, emerging from the pupal stage with a full complement of eggs, which they deposit as quickly as possible. Adults are capable of flight but tend to run rapidly over the leaf surface making short flights or jumps from leaf to leaf.

This species is commercially available in many countries.

- *Life cycle:* Leafhoppers insert their eggs into secondary veins on the undersides of leaves and the parasitoid lays its egg into the eggs of the leafhopper. Initially nothing is visible, as the host egg is transparent, but after a few days the minute red eyes of the leafhopper can be determined, soon after which the egg turns red as the *A. atomus* develops (Figure 4.249). When fully mature, the parasitized leafhopper egg changes to a darker red/brown color from which the adult *A. atomus* emerges. The whole life cycle from egg to adult takes some 12 to 20 days at 25°C to 16°C (77°F to 60°F), respectively.

- *Crop/pest associations:* The parasitoid occurs quite frequently among leafhopper colonies and can build up to high numbers, giving a high level of control. However, this usually occurs late in the growing season after considerable damage has been caused by the pest. It should also be remembered that being an egg parasitoid, any unparasitized eggs that survive will pass through all the developmental stages before reaching adulthood to provide the appropriate stage for further parasitism.

- *Influence of growing practices:* It is known that many chemicals will kill the adults, but all their development occurs within the host egg, so they are protected from most pesticides. Therefore, short persistence and selective compounds integrate reasonably well with *A. atomus*.

Family: Aphelinidae

APHELINUS ABDOMINALIS

The adult wasp is black and yellow, 3 mm in length, and prefers to walk over the crop rather than fly (Figure 4.250). This habit tends to make the parasitoid remain within the crop. Although it reproduces as a parasitoid, host feeding is an important source of nutrition for the adult wasp that effectively increases aphid

Figure 4.250   Adult parasitoid wasp (*Aphelinus abdominalis*) with ovipositor inserted in potato aphid (*Macrosiphum euphorbia*); note wings folded back over body, revealing ovipositor.

Figure 4.252   Adult *Aphelinus abdominalis* with wings folded back showing ovipositor inserted in potato aphid (*Macrosiphum euphorbia*).

Figure 4.251   Adult parasitoid wasp and black mummy of *Aphelinus abdominalis* with healthy potato aphids (*Macrosiphum euphorbia*); note the wings flat against body.

Figure 4.253   Black mummies of *Aphelinus abdominalis* showing exit hole cut by adult wasp prior to emergence.

mortality, with each female killing approximately two aphids per day. Parasitized aphids turn black when the parasitoid pupates within the host body and the "mummified" bodies remain attached to the leaf (Figure 4.251).

This species is commercially available in many countries.

- *Life cycle:* Female *Aphelinus abdominalis* select an aphid of an appropriate size and species before reversing up to it with wing tips lifted and the ovipositor extended; the tip is inserted into the ventral surface and an egg laid (Figure 4.252). This hatches within the aphid and the larva feeds on the host's internal tissues. While this

activity proceeds, the aphid continues to feed and grow, and it may produce some offspring before finally being killed as vital body organs are consumed just before parasitoid pupation. At 20°C (68°F), a black mummy forms 7 days after parasitism, and some 14 days later the adult cuts a circular hole in the aphid and emerges (Figure 4.253).

- *Crop/pest associations:* The principal host aphids of protected crops are *Macrosiphum euphorbiae* and *Aulacorthum solani* on which *A. abdominalis* prefers to parasitize the second and third instar nymphs. Larger aphids are less frequently attacked, while first and small second instar nymphs are used as food by the adults. Fecundity is low during the first few days of adult life, but by the fourth day may rise to 10–15 eggs/day and, providing host predation continues, does not decline with age. Adult females have a longevity of between 15 and 27 days and may parasitize over 200 aphids, plus kill 40 or more in the process of host feeding.

- *Influence of growing practices:* *A. abdominalis* has been recorded as parasitizing a wide range of aphid species in both protected and outdoor crops. Developing wasps still in the aphid skin (mummy stage) are protected from most short-persistence insecticides such as soap-based products and plant-extract oils that would otherwise kill the adult on contact. Commercial supplies are distributed as black mummies in a vial from which the adults emerge some 2–3 days after receipt. Mummies should be placed directly on the aphid infestation due to host predation by adult wasps. Although they have a wide host range, other parasitoids, such as *Aphidius* spp. may be more economical to use against aphids such as *Myzus persicae* and *Aphis gossypii*.

Figure 4.254   Adult parasitoid wasp (*Encarsia formosa*) among whitefly larval "scales."

ENCARSIA FORMOSA

The whitefly parasitoid *Encarsia formosa* (Figure 4.254) was found in England and successfully used first in 1926. Within 2 years, 250,000 parasitoids had been reared for use on nurseries around England, France, and then on to Canada. Adult females are 0.6 mm long with a black head and thorax, a yellow abdomen, and translucent wings. *E. formosa* males are of similar size to females, but are all black and usually only make up about 1%–2% of the population. They may be produced when a second egg is laid into a host already parasitized by *E. formosa*. This is known as autoparasitism and occurs most frequently at very high temperatures. The most obvious sign of *Encarsia* spp. activity is the presence of black "scales" on leaves. These are the pupal stages of the parasitoid

and are formed inside the pupae of the whitefly.

This species is commercially available in many countries.

- *Life cycle:* Adult wasps are attracted to the host whitefly "scale" (so called because the larval stage of whitefly is mostly immobile and resembles miniature scale insects) by volatile compounds given off from whitefly honeydew (see p. 42). Adults feed on honeydew and also from wounds to the whitefly larvae made by their ovipositor. This can kill some scales, so neither whitefly nor parasitoid develops; this is known as host feeding (see figure in next section). Usually, a single egg is laid that passes through three larval stages, during which time the whitefly scale remains white and develops normally. When fully developed (about 10 days after oviposition at 22°C [72°F]) the whitefly scale turns black as the parasitoid pupates. The pupae remain attached to the leaf and the adult emerges some 10 days later from a hole cut through the puparium with a special "tooth" (Figures 4.255 and 4.256).
- *Crop/pest associations:* *E. formosa* is introduced to crops as black scales from which adults emerge a few days later. The majority of these wasps are mass

(see p. 42)

ORDER: HYMENOPTERA (PARASITOID WASPS)

Figure 4.255 *Encarsia formosa* emerging from parasitized black "scale."

Figure 4.257 Nicotiana leaf with parasitized (black) glasshouse whitefly (*Trialeurodes vaporariorum*) as used for commercial mass production of this parasitoid wasp.

Figure 4.256 Adult *Encarsia formosa* recently emerged from parasitized black "scale."

Figure 4.258 Close-up of leaf with empty, gray/white pupal cases of *Trialeurodes vaporariorum* and parasitized black "scales" of *Encarsia formosa*; brown "scales" are signs of host feeding by the adult parasitoid wasps.

produced on large, flat tobacco leaves *Nicotiana tabacum* (Figures 4.257 and 4.258) from which the mature black scales may be removed, graded, and stuck to cards for distribution within a crop (see Figure 1.7 in Chapter 1). The safe emergence of this parasitoid is characterized by a D-shaped exit hole in the top of the scale, which can clearly be seen by shining a light through from behind the leaf. Under ideal conditions, up to 300 eggs can be laid at a rate of approximately 12–15 per day. The structure of the host plant influences *E. formosa* parasitism. For instance, cucumber has hairy leaves that trap whitefly honeydew, making mobility difficult; this is not the case on aubergine, tomato, and most ornamentals unless the plants are heavily infested with whitefly.

- *Influence of growing practices:*
E. formosa is commercially available
throughout the year, but is most active
at above 18°C (64°F) with good light
intensity. Long periods of poor light
and cooler conditions disrupt its activity
and can allow the whitefly to become
troublesome. Adult whitefly lay their
eggs on the apical leaves, and as the
plant continues growing upward, the
developing scales tend to be lower down
the plant. On crops such as tomato,
where the lower leaves are regularly
removed to allow light and access to
the fruit for picking, deleafing too high
up the plant can deplete the parasitoid
black scales before they emerge.

ENCARSIA TRICOLOR

These small parasitoid wasps of whitefly are
slightly larger at 0.8 mm than the closely
related *E. formosa*. They are reported to
occur throughout much of Eurasia; however, they are more commonly found
in the cooler parts of Europe from the
Mediterranean to Scandinavia. Adult males
and females have a yellow head and antennae and a gray-yellow thorax with a lighter
yellow and black abdomen; wings are transparent (Figure 4.259a). Usually, the first
sign of the wasps' presence on plants are
black "scales" on leaves—the pupal stage of
parasitized whitefly (Figure 4.259b). *E. tricolor* has been collected from several species
of whitefly and on various crops including
glasshouse whitefly (*Trialeurodes vaporariorum*), tobacco whitefly (*Bemisia tabaci*),
and cabbage whitefly *Aleyrodes proletella*.

- *Life cycle:* Adult wasps are attracted
to whitefly larval scales by the volatile
compounds released from whitefly
honeydew. Eggs are laid directly
into second, third, and fourth instar
larvae as well as early stage pupae
(Figure 4.259c) (see whitefly life
cycle, p. 42) at a rate of up to 5–7 per
day for up to 3 weeks. Eggs are laid

(a)

(b)

(c)

Figure 4.259 (a) Adult *Encarsia tricolor* **among
larval stages of cabbage whitefly (***Aleyrodes
proletella***). (b) Black pupal stages of cabbage
whitefly (***Aleyrodes proletella***) parasitized by**
Encarsia tricolor; **white "scales" may be unparasitized or may contain a parasitoid egg or young
larva. (c) Adult female** *Encarsia tricolor* **ovipositing into a fourth instar larval stage of** *Aleyrodes
proletella.*

singly and hatch to a minute larva
that develops entirely inside the host.
Parasitized whitefly scales continue
to develop through to pupation, at
which time the scale turns black as

ORDER: HYMENOPTERA (PARASITOID WASPS)

the parasitoid pupates. Males develop as a hyperparasitoid when a female *E. tricolor* lays an egg into a young parasitoid larva of the same or another related species; the second parasite kills the original parasite. *E. tricolor* are also recorded as host feeding whereby the ovipositor is inserted several times and in a saw-like movement the internal organs sliced; the adult then feeds on the weeping body fluids. Whitefly pupae remain attached to the host plant leaf; healthy (unparasitized) cabbage whitefly (*A. proletella*) are gray (summer generation) to brown (winter generation), while black indicates successful parasitization. The presence of a T-shaped emergence hole in the top of the pupal scale is from an adult whitefly, while a circular hole shows that an adult wasp has emerged.

- *Crop/pest associations:* There is much interest in its ability to parasitize cabbage whitefly as well as other species at lower temperatures, particularly commercial protected crops, where any reduction in heating is considered a financial saving. Augmentative releases, in association with insect-proof netting to reduce migrating adult whitefly, are being studied to improve whitefly control in temperate field crops. Wider use depends on the economic mass production of parasitoids, either on the primary target host or an alternative host that may be easier to rear.
- *Influence of growing practices:* These naturally occurring wasps are being recorded in areas of reduced pesticide intervention and, due to the continuing threat of pesticide resistance, there is considerable interest in their wider commercial use. The effects of male production by hyperparasitism have been studied in laboratory trials and are regarded as a population regulation

mechanism, but the full effects in a field population are debatable.

ERETMOCERUS EREMICUS

Eretmocerus eremicus is a small (0.5–0.6 mm) ectoparasitoid wasp that attacks the larval stages of whitefly, in particular *Trialeurodes vaporariorum* (the glasshouse whitefly) and *Bemisia tabaci* (the tobacco or sweet potato whitefly) (Figure 4.260). The latter species has a notifiable status in several countries due to its potential to transmit plant viruses of the geminivirus group, which includes tobacco leaf curl and tomato yellow leaf curl. These viruses can affect a wide range of hosts, particularly in the Solanaceae. *E. eremicus* originates from the deserts of Arizona and California, where it tolerates extremes of temperature from almost freezing to over 40°C (104°F).

This species is commercially available in many countries, but a license for its release is required in some.

- *Life cycle:* Female parasitoids deposit a single egg under the whitefly larval scale (Figure 4.261) that, on hatching, burrows into the host, where a protective capsule is formed around the developing parasitoid. Although eggs can be laid beneath any of the larval instars, there is a preference for the second instar.

Figure 4.260   Adult parasitoid wasp (*Eretmocerus eremicus*) with cotton whitefly (*Bemisia tabaci*) larval "scale."

Figure 4.263   Glasshouse whitefly (*Trialeurodes vaporariorum*) parasitized by *Eretmocerus eremicus* showing exit hole cut by adult when emerging; note that the whitefly "scales" turn creamy yellow when parasitized.

Figure 4.261   Adult parasitoid wasp (*Eretmocerus eremicus*) ovipositing in tobacco or sweet potato whitefly (*Bemisia tabaci*) larval "scale."

Figure 4.262   Pair of adult *Eretmocerus eremicus* among parasitized and unparasitized *Trialeurodes vaporariorum* whitefly "scales."

Successful parasitism shows as a yellow color and visible presence of a parasitoid body (Figure 4.262). Close to emergence, the body and eyes of the adult can be clearly seen through the dead whitefly cuticle: The adult wasp cuts an exit hole through which it emerges (Figure 4.263). At an average temperature of 23°C (74°F), parasitoid development from egg to adult takes 17–18 days, with the pupal stage occupying up to 8 days of this time. The life cycle slows at lower temperatures, taking some 40 days at 17°C (63°F).

- *Crop/pest associations: Eretmocerus* spp. are becoming extremely useful for whitefly control when continuous high temperatures adversely affect *Encarsia formosa* and also for crops where there is a threat or presence of *Bemisia tabaci*. Adult wasps are introduced to the crop at rates of up to 20/m^2 in areas of heavy whitefly infestation for up to 6 weeks, or until adequate control is achieved. Where pest levels are lower or additional control is being used, rates of three to five per square meter will provide reasonable protection from the major whitefly species.

- *Influence of growing practices:* Although the wasp is not as efficient as *Encarsia formosa* in terms of parasitoid reproduction, *E. eremicus* can result in equally high numbers of whitefly mortality due to host feeding. The overall result is comparable in terms of pest control for temperate environments, but exceeds *E. formosa*, where high temperatures and low humidities are common.

Family: Braconidae

APHIDIUS COLEMANI

Aphidius colemani is a small, black wasp, 4–5 mm long, that inserts a single egg into a host aphid (Figure 4.264). All other life stages are within the aphid. The appearance of a golden brown mummy indicates the presence of these parasitoids on a crop (Figure 4.265). Principal aphids attacked by *A. colemani* include *Aphis gossypii* (melon or cotton aphid), *Myzus persicae* (the peach–potato aphid), and *M. nicotiana* (the tobacco aphid). They are also recorded feeding on other aphids, but less information is available. In general this parasitoid attacks the smaller aphid species.

Figure 4.264   Adult parasitoid wasp (*Aphidius colemani*) ovipositing in young peach–potato aphid (*Myzus persicae*).

This species is commercially available in many countries.

- *Life cycle:* The life cycle is similar to that of *Aphidius ervi*, except that the cycle of *A. colemani* is shorter (for details, see next entry), taking only 14 days at 21°C (70°F) to reach adulthood and approximately 20 days at 15°C (59°F), compared to nearly 19 and 29 days, respectively, for *A. ervi* at the same temperatures.

- *Crop/pest associations: Aphidius* spp. are good at host location and can provide reasonable levels of control if introduced early when pest numbers are low. However, if aphids are established in colonies, *A. colemani* will take some time to make an impact on the pest population, so predators or a selective pesticide should be considered.

- *Influence of growing practices:* Prolonged temperatures above 30°C (86°F) may reduce the efficacy of this parasitoid, particularly on protected crops, which should be monitored during hot weather and treated with a compatible insecticide if necessary. The mummy stage is tolerant to most short-persistence pesticides, but those such as synthetic pyrethroids have long residual activity and may kill the adult as it emerges from the aphid mummy (Figure 4.266). Banker plants of cereals

Figure 4.265   Parasitized mummies of *Aphidius colemani* among colony of cotton aphid (*Aphis gossypii*) on cucumber leaf.

Figure 4.266   Close-up of parasitized mummies of *Aphidius colemani* showing exit holes from cotton aphid (*Aphis gossypii*) on cucumber leaf.

infested with a specific aphid are useful in crops where a continuous supply of parasitoids is required.

APHIDIUS ERVI

Aphidius ervi is of European origin, but has been widely introduced into Australia, North and South America, and other countries. Adults are black and 4–5 mm in length (Figure 4.267). *A. ervi* tends to parasitize larger aphid species such as *Macrosiphum euphorbiae* (the potato aphid), *Aulacorthum solani* (the glasshouse potato aphid), and *Acyrthosiphon pisum* (the pea aphid) and it is these aphids that are targeted commercially. Several other aphid species are also attacked.

This species is commercially available in many countries.

- *Life cycle:* In most respects the life cycle of *A. ervi* is very similar to that of *A. colemani* (see previous entry). The adult female locates a single aphid or a small colony and examines an individual by palpating it with her antennae to determine the size and if it has already been parasitized. If a suitable host is found, the wasp rapidly curls her abdomen under her body and stabs the aphid with her ovipositor to insert an egg (Figure 4.268) that hatches a few days later. The parasitoid larva

interferes little with the development or outward appearance of the aphid, except that feeding increases and the aphid excretes more honeydew. Affected aphids can produce healthy offspring. When fully developed, the parasitoid larva cuts a slit in the underside of the hollowed-out aphid and attaches the carcass to the leaf or other surfaces including flowers, fruits, pots, or string. A silken cocoon is spun within the host body, which takes on a golden brown color as the parasitoid pupates and forms the characteristic aphid mummy (Figure 4.269). After 5–10 days an adult has developed and cuts a circular door in the back of the mummy through which it leaves to seek out new hosts.

Figure 4.268 Parasitoid wasp (*Aphidius ervi*) ovipositing into glasshouse potato aphid (*Aulacorthum solani*); note droplets of alarm pheromone produced by the aphid being attacked.

Figure 4.267 Parasitoid wasp (*Aphidius ervi*) getting ready to oviposit into glasshouse potato aphid (*Aulacorthum solani*).

Figure 4.269 Pea aphids (*Acyrthosiphon pisum*) parasitized by *Aphidius ervi*: mummy stage.

- *Crop/pest associations:* Pest control by parasitoids is initially slow, but as long as a degree of damage can be tolerated, it is possible to achieve 100% control. Under these conditions, swarms of adults can be seen flying close to the tops of plants hunting for aphids to parasitize. A banker plant system may be used to produce high numbers of these parasitoids by infesting cereals with a host plant-specific aphid.
- *Influence of growing practices:* A. ervi is active in most crop situations where the above aphids are present. Its efficacy may be reduced above 30°C (86°F) and below 8°C–10°C (46°F–50°F), although in the mummy stage it is able to survive frosts. The mummy is also tolerant of most short-persistence insecticides (soft soap and natural insecticides), but contact or residual activity of other chemicals can kill adults. *Aphidius* spp. may suffer from hyperparasitoids, which can reduce the level of control achieved, particularly late in the season (see previous Figures 4.246 and 4.247).

COTESIA GLOMERATA (SYN. APANTELES GLOMERATUS)

Cotesia glomerata adults are about 7 mm in length with long (1.5 mm), upwardly curving antennae (Figure 4.270). *C. glomerata* is commonly found throughout most of Europe and North America, where it was introduced with great success to control both large and small cabbage white butterfly larvae. The characteristic sulfur-yellow silken cocoons form in irregular masses on almost any surface, including leaves and stems in summer and walls and roofs of buildings during winter.

- *Life cycle:* Females mate immediately after emerging from their pupal cocoon (Figure 4.271) and start laying eggs very soon after. Each female can deposit 20–40 eggs per host and a total of up to 200 eggs can be laid per adult. They prefer first or young second instar caterpillars of Pieridae butterflies and, after oviposition, eggs remain dormant until the host larvae is nearly fully grown. Young *C. glomerata* larvae devour the internal tissues of the living caterpillar; after 15–20 days, when they

Figure 4.271   Adult *Cotesia glomerata* with transparent empty pupal cocoon from which it has emerged.

Figure 4.270   Adult *Cotesia glomerata* recently hatched from parasitized caterpillar body.

are fully developed, they rupture the host body to spill out (Figure 4.272), forming a mass of individual silken cocoons, often enveloped by a common web (Figure 4.273). The life cycle takes 22–30 days from egg to adult depending on temperature. They overwinter within the cocoons, awaiting the next spring before emerging.

- *Crop/pest associations:* Levels of parasitism during early spring are usually low, but these can rise to almost 75% by late summer, making an extremely useful contribution to caterpillar control on many brassica crops. *C. glomerata* has been shown to transmit the granulovirus of the small white butterfly, *Pieris rapae*,

leading to even higher levels of pest control.

- *Influence on growing practices:* The larvae and adults of most *Cotesia* species are sensitive to pesticides, and even Lepidoptera-specific products such as *Bacillus thuringiensis* may kill developing larvae unless they can leave the caterpillar before it dies. Adult wasps will survive the bacterium, provided sufficient healthy hosts also survive. This braconid wasp is not commonly attacked by many of the hyperparasitoid wasps. This species is not commercially produced as a biocontrol agent because it only manifests itself after the caterpillar is fully grown and the host plant severely damaged.

DACNUSA SIBIRICA

Adult *Dacnusa sibirica* (Figure 4.274) are black, 2.5–3 mm in length, with long antennae (about the same as length of the body) and can easily be mistaken for aphid parasitoids (*Aphidius* spp.). This wasp is a koinobiont endoparasitoid and lays its egg directly into a young leaf miner larva, without killing it (see p. 2). The parasitoid larva is light gray in color and curled in a C-shape. It develops with the host larva and pupates within the leaf miner puparium.

Figure 4.272   Larvae of *Cotesia glomerata* emerging from parasitized cabbage white butterfly caterpillar (*Pieris brassicae*).

Figure 4.273   Parasitized cabbage white butterfly caterpillar (*P. brassicae*) showing sulfur yellow cocoons of *Cotesia glomerata*.

Figure 4.274   Adult parasitoid wasps (*Dacnusa sibirica*) in courtship dance on leaf containing leaf miner larva.

ORDER: HYMENOPTERA (PARASITOID WASPS)

This species is commercially available in many countries.

- *Life cycle:* Host location by the adult parasitoid is initially by scent of the leaf miner frass within the damaged leaf tissue. When a promising leaf has been found, the mining larva is located by antennal drumming and a single egg laid inside the host larva, which continues to develop through to pupation (Figure 4.275). The egg and larva of *D. sibirica* can only be found by dissection of the host leaf miner larva. Similarly, during the early stages of pupation it is difficult to distinguish between healthy and parasitized insects, although parasitized pupae tend to be dark gray/black (Figure 4.276)

Figure 4.275   Adult *Dacnusa sibirica* ovipositing into leaf miner larva within leaf.

Figure 4.276   Chrysanthemum leaf miner (*Phytomyza syngenesiae*) pupa excavated from leaf to show color change to gray/black when parasitized by *Dacnusa sibirica*.

instead of creamy white prior to adult emergence. However, as the parasitoid emerges some 3–5 days before healthy, nonparasitized leaf miners, collection of pupae to monitor adult emergence gives a good indication of the parasitoid efficacy. This is also a useful method of collecting parasitoids.

- *Crop/pest associations:* This parasitoid works well at low pest densities and may be introduced early in the growing season. It is particularly useful in tomato crops when released during late winter and early spring. The parasitoid *Diglyphus isaea* is favored as the season progresses and higher numbers of leaf miners may be found. Use of *D. sibirica* on other commercial crops is dictated by the amount of leaf damage that can be tolerated before adequate control is achieved.

- *Influence of growing practices:* The development time of the parasitoid is some 3–5 days shorter than the host insect and, with adult females able to lay over 150 eggs each, they can make a reasonable impact on a pest infestation. Temperatures above 22°C (72°F) result in a reduction of egg production. Although *D. sibirica* is a useful and persistent parasitoid against a wide range of leaf miner species, it cannot be regarded as a total control agent and is not as effective as *Diglyphus* spp.

DIAERETIELLA RAPAE

The widely distributed parasitoid *Diaeretiella rapae* is found principally attacking aphids on cabbage, potato, and sugar beet, but is also able to parasitize several other species found on related host plants. Adults are 4–5 mm in length, a shiny, dark brown color, and may be found feeding on honeydew while they search for aphids (Figure 4.277). Parasitized aphids take on a light yellowish color and have a papery appearance to their skin, almost like an inflated paper bag.

Figure 4.277 Adult parasitoid wasp (*Diaeretiella rapae*) ovipositing into mealy cabbage aphid (*Brevicoryne brassicae*).

Figure 4.278 *Brevicoryne brassicae* mummies parasitized by *Diaeretiella rapae;* the lower one shows the exit hole where the adult wasp has emerged.

- *Life cycle:* Single aphids or small colonies are located by female *D. rapae*. Individual aphids are examined by antennal palpation to determine size and evidence of previous parasitism. After a suitable host has been selected, the female curls her abdomen under her body and holds the aphid with her antennae. An egg is inserted, which hatches after a few days. The parasitized aphid continues to develop quite normally except that slightly more honeydew is produced. When the parasitoid larva is fully developed, it cuts a slit along the underside of the now empty aphid body and attaches the carcass to the leaf or other surface by spinning a silken cocoon in which it pupates. After 6–12 days the adult has developed within the mummy and cuts an almost circular hole in the back of the aphid, through which it leaves (Figure 4.278).
- *Crop/pest associations:* Three or more generations of parasitoids may occur each year and a considerable number of aphids can be parasitized. However, due to the dense colonies of aphids such as *Brevicoryne brassicae*, effective control is rarely achieved. Higher numbers of parasitoids have been found when areas around cultivated fields have been allowed to grow a variety of wild plants such as *Chenopodium* (fat hen), thistles, and brassica plants.
- *Influence of growing practices:* D. rapae is sensitive to many pesticides used in commercial agriculture, and numbers may be depleted in areas of intensive cropping. The use of selective pesticides compatible with ICM techniques helps to increase parasitoid populations. Unfortunately, this may have detrimental effect when marketing a crop due to high numbers of mummies on plants, as happened in the UK in 1998, resulting in many cauliflower heads being downgraded.

PRAON MYZIPHAGUM AND *PRAON VOLUCRE*
Praon myziphagum and *P. volucre* are polyphagous parasitoids that frequently attack winged aphids, which in turn carry the immature parasitoid long distances during their migratory flights. Adults are 2.5–3 mm in length, with a shiny black head and thorax and brown abdomen and legs (Figure 4.279). *Praon* spp. are distinctive in that the last larval instar and pupa are in a silvery-tan silk pad spun beneath the mummified body of the host aphid. Adults feed on honeydew and are not recorded as killing aphids by host feeding.

ORDER: HYMENOPTERA (PARASITOID WASPS)

Figure 4.279 Adult parasitoid wasp (*Praon myziphagum*) searching among aphids for suitable host.

Figure 4.281 Mummified alate (winged) aphid killed by *Praon myziphagum*.

Figure 4.280 Adult *Praon myziphagum* ovipositing in glasshouse potato aphid (*Aulacorthum solani*).

This species is commercially available in many countries.

- *Life cycle:* Female *Praon* spp. lay most of their eggs within the first 4–5 days after emergence. They select suitable second- and third-stage nymph aphids with their antennae and, once located, insert a single egg (Figure 4.280) that hatches after 3–4 days. Parasitized aphids continue to develop and on reaching maturity can reproduce until quite late in the parasitoid's life cycle. The last larval instar cuts a slit in the underside of the host and spins a double-walled silken cocoon that raises the aphid body off the leaf surface (Figure 4.281). Adult emergence occurs after 2–3 weeks at 16°C–24°C (61°F–75°F) when a small exit hole is bitten through the side of the raised silken cocoon.

- *Crop/pest associations: Praon* spp. are commonly found in the spring as the first naturally occurring aphid parasitoids. Their wide host range makes them useful for inoculative introductions on most crops. As with other aphid parasitoids, initial control may be slow to achieve but their presence is usually welcome.

- *Influence of growing practices:* For successful adult emergence, a reasonably high humidity is required (average protected cropping conditions are generally adequate), which may preclude use in interior landscaping. The presence of too many silvery mummies on leaves of ornamentals may affect the quality appearance of the product. Parasitoids within the mummy cocoon are tolerant of most short-persistence pesticides but can be killed by broad-spectrum and long-persistence insecticides.

Family: Ichneumonidae
DIADEGMA INSULARE

Principally a parasitoid of diamond-back moth larvae (*Plutella xylostella*), the Ichneumonid wasp *Diadegma insulare* is found throughout Europe and North America. Other species of *Diadegma* may also be found in Asia. Adults are 6 mm long and may be seen roaming over crop foliage

**Figure 4.282   Adult ichneumonid wasp
(*Diadegma semiclausum*).**

searching for host larvae (Figure 4.282) in which to lay an egg. They are also to be found in or on flowers, where they feed on nectar.

- *Life cycle:* Females mate immediately after emerging from their pupal cocoon and then start searching for a suitable young host larva by drumming their antennae. Once located, the wasp curls her abdomen under her body and injects a single egg into the moth larva with her ovipositor. The caterpillar is not paralyzed and continues to develop through to pupation, at which time the parasitoid exits the host cocoon to spin its own within that of the host. The cycle takes 12–18 days with up to six generations each year. Overwintering takes place in crop debris as a pupa within the host cocoon.
- *Crop/pest associations:* As with most parasitoids, activity can be slow during spring but rapidly increases to 70%–90% efficacy by late summer, provided adequate food sources are available. Due to its widespread habitat, *D. insulare* is thought to attack other similar host species of caterpillar that feed on brassica crops.
- *Influence of growing practices:* Adult longevity and fecundity are both increased when nectar is available, as other sources of sugars, such as from insect honeydew, are reported to be

inferior to plant nectar. The presence of nectar-producing plants either as crops or as flowering weeds is thus vitally important for the success of this parasitoid. *D. insulare* adults are susceptible to most insecticides. Developing larvae can also be killed if pesticides, including the Lepidoptera-specific *Bacillus thuringiensis*, affect the host caterpillar.

Family: Eulophidae

DIGLYPHUS ISAEA

This is an idiobiont ectoparasitoid that paralyzes its host before laying one or more eggs next to the larva (see p. 2). Adults are small and vary in length from 1 to 2.3 mm. They have a black background color with a metallic green sheen, and females have a yellow stripe on the hind leg (Figure 4.283). *D. isaea* is common throughout the summer on leaf miner-infested weeds, particularly milk thistles (*Sonchus arvensis*), which can produce up to 80 parasitoids per plant, making them worth collecting to harvest the parasitoids.

This species is commercially available in most countries.

- *Life cycle:* Adult females generally attack second instar leaf miner larvae (mines approximately 2 cm in length).

**Figure 4.283   Pair of adult parasitoid wasps
(*Diglyphus isaea*) investigating a leaf miner larva within the leaf.**

Host location is by antennal drumming and the scent produced by the feeding leaf miner. Once a suitable host has been found, the wasp stings the larva (Figure 4.284), permanently paralyzing it before depositing one to three eggs immediately next to it (Figure 4.285). The parasitoid larvae insert their mouthparts into their host and develop a blue green color as they reach maturity (Figure 4.286). When fully grown the parasitoid larvae move away from the dead host and collect frass pellets, which are placed one above the other to produce miniature "pit props" (Figure 4.287). These prevent damage to the pupae as they develop to adulthood before emerging from the leaf through a hole in the epidermis.

- *Crop/pest associations:*
These commercially available parasitoids work better at high or increasing pest densities and may be introduced when leaf miner numbers reach one new mine per plant per week. They are particularly useful in tomato crops when released during early spring. The parasitoid requires sunny conditions for maximum activity and frequently produces more than one parasitoid from each host larva. Where multiparasitism occurs, the resulting adults vary greatly in size. *D. isaea* usually kill more leaf miners

Figure 4.285   Young *Diglyphus isaea* parasitoid wasp larva next to the larger leaf miner larva excavated from tunnel in leaf.

Figure 4.286   Mature larva of *Diglyphus isaea* excavated from leaf miner tunnel; note black frass pellets used to create "pit-props."

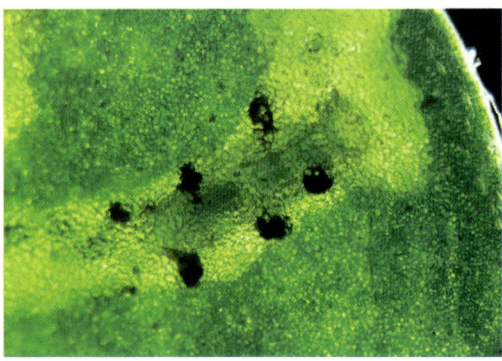

Figure 4.287   "Pit-props" inside leaf miner tunnel created by the parasitoid *Diglyphus isaea* larvae prior to pupation.

by stinging and feeding on the wound than by parasitism, particularly when the wasps are young and eggs are still developing. The ability of *D. isaea* to multiply rapidly and to hyperparasitize

Figure 4.284   Adult *Diglyphus isaea* ovipositing into a leaf miner larva.

Dacnusa sibirica and other leaf miner parasitoids usually makes it the dominant species of leaf miner parasitoid during summer months.

- *Influence on growing practices:* Under conditions of low light, too few leaf miners, or hosts of the wrong age range, *D. isaea* may be slow to exert its full control potential. However, once fully established in a crop, the results can be spectacular by a combination of multiparasitism and host feeding by adult wasps on mining larvae. Although most acaricides and fungicides can be integrated with these parasitoids, it is thought that the use of sulfur against powdery mildew as a fumigant or spray disrupts host location by the parasitoid.

TETRASTICHUS ASPARAGI (SYN. *T. COERULEUS*)
Tetrastichus asparagi is an egg parasitoid of the common asparagus beetle *Crioceris asparagi*. It is also likely to attack the 12-spotted asparagus beetle *C. duodecimpunctata*. Adults are dark metallic green in color and 1.5 to 2 mm in length; their pinched waist allows greater articulation of the abdomen during oviposition (Figure 4.288).

- *Life cycle:* Adult female wasps drum over the plant surface with their antennae to locate potential eggs. Once located, further drumming is done on the egg to determine its status for either egg-laying or host-feeding (Figure 4.289). A sharp, needle-like ovipositor is drilled into the host egg; young eggs are preferred for host feeding, older eggs containing an embryo or young larva are required for oviposition, and preparasitized eggs are rejected. The parasitoid egg hatches to a minute larva that feeds on the developing host larva, passing through all stages before emerging as an adult 2–3 weeks later.

Figure 4.288   Adult parasitoid wasp (*Tetrastichus asparagi*) with asparagus beetle (*Crioceris asparagi*) egg.

Figure 4.289   Adult parasitoid wasp (*Tetrastichus asparagi*) with asparagus beetle (*Crioceris asparagi*) eggs.

- *Crop/pest associations:* They are found in most asparagus-growing regions of the world and, like the beetle on which they prey, are host plant-specific. Many reports estimate up to 70% control of *C. asparagi* occurring in most years. However, because it is an egg parasitoid, the dark brown egg case remains attached to the plant after the adult wasp has emerged; this can lead to commercial downgrading if pest numbers are high.
- *Influence of growing practices:* Early season control of asparagus beetle is one of the most important factors for commercial producers. The use of short-persistence insecticides such as natural pyrethrum, which breaks down within a few days, integrates

reasonably well, allowing a population of *T. asparagi* to establish and maintain control through the plants' regeneration season.

Family: Encyrtidae

ENCYRTUS INFELIX

Adult *Encyrtus infelix* are black, about 2 to 2.5 mm in length; adult size depends on the size and nutritional value of the host scale insect. They are parasitoids of soft scale insects such as *Saisettia coffeae*. This is an all-female species (thelytokous), so no mating is required. Eggs are laid into imma-ture adult scale insects that, at the time of parasitization, are large and creamy gray in appearance (Figures 4.290 and 4.291). Mature scales are brown in color and too old for parasitism.

Figure 4.290   Adult *Encyrtus infelix* searching for suitable host stage of the soft scale insect *Saisettia coffeae* for egg laying.

Figure 4.291   Adult *Encyrtus infelix* feeding on droplet of scale insect honeydew.

This species in commercially available in many countries.

- *Life cycle:* Females select healthy, creamy white colored scale insects and insert a single egg through the ventral cuticle into the host's body. This hatches to a minute larva that feeds internally to kill the scale and prevent any eggs maturing. As the parasitoid develops, its host changes color to a darker gray-black when the grub pupates. Adults emerge by cutting a roughly circular hole through the host dorsal cuticle (Figure 4.292). Generation time is 3 to 4 weeks at 25°C (77°F). At this temperature each adult can lay approximately 30 or more eggs. The host scale insect's life cycle is closer to 2 months at the same temperature. Adult females generally contain fully developed eggs at the time of emergence and can begin oviposition almost immediately after they have emerged. High temperatures (above 35°C [95°F]) may initiate ovipositional diapauses where fewer, in any, eggs are laid.

- *Crop/pest associations:* This parasitoid works well at low pest densities and may be introduced early in the growing season. For best control of soft scale insects, it should be used in conjunction with *Metaphycus helvolus* (see p. 178).

Figure 4.292   Parasitized soft scale insect (*Saisettia coffeae*) showing circular exit hole produced when adult *Encyrtus infelix* emerged.

This is because of the age-specific requirement of *E. infelix* compared to *M. helvolus* (see image of adult parasitoid wasp on page 178), which attacks and feeds on very young scale insects and then lays its eggs into slightly larger immature scales; however, they cannot develop in older ones.

- *Influence of growing practices:* The development time of the parasitoid is some 3–5 days shorter than that of the host insect; because all adults are female and lay over 30 eggs each, they can make a reasonable impact on a pest infestation. Generally, a single introduction is sufficient to establish the parasitoid, providing that the scale insect population has sufficient numbers of mixed-age hosts. In a continually warm environment, they may go from year to year with no further introductions. The Brazilian plume flower (*Jacobinia carnea*) makes an excellent banker plant for both scale insects and their parasitoids, whereas scale insect control on most other plants can be almost complete. These (and most other beneficials) are extremely sensitive to insecticides, particularly synthetic neonicotinoids, for which at least 1 year should be allowed between the last application and introduction of the beneficials.

Figure 4.293   Adult parasitoid wasp (*Leptomastix dactylopii*).

Figure 4.294   Adult parasitoid wasp (*Leptomastix dactylopii*) searching over citrus mealybug (*Planoccus citri*) eggs and young nymphs.

LEPTOMASTIX DACTYLOPII AND *LEPTOMASTIX EPONA*

These small (about 3 mm long) koinobiont endoparasitoid wasps (see p. 2) are a useful addition to the control of the mealybug *Planoccus citri* and *Pseudococcus viburni* (the citrus and glasshouse mealybugs, respectively). *L. dactylopii* is yellowish-brown in color with relatively long, hinged antenna (Figure 4.293), which it taps to locate the host (Figure 4.294). *L. epona* is more uniformly dark brown in color. Adult wasps are good fliers and with their hopping flight can provide excellent control of mealybugs even when the host is at low density.

This species is commercially available in many countries, but a license for its release is required in some.

- *Life cycle:* The female lays between 80 and 100 eggs, singly, into third instar

Figure 4.295   Adult parasitoid wasp (*Leptomastix dactylopii*) searching over citrus mealybug (*P. citri*).

and/or adult mealybugs (Figure 4.295). The developing wasp larva feeds on the host body fluids, consuming the majority of the mealybug. Where the host provides insufficient nutrition due to its size or age, male parasitoids are usually produced, but in general the sex ratio is 1:1 with unmated females producing only male offspring. Pupation occurs within the mummified skin of the host, which resembles a slightly bulbous body covered with fine threads. When fully developed, the parasitoid cuts a hinged lid through the mealybug skin and pushes its way out. Egg-to-adult development takes 3–4 weeks depending on temperature.

- *Crop/pest associations: L. dactylopii* is specific against *P. citri*, which can be identified by its very short tail filaments and darker bar along its back, while *L. epona* will attack *P. viburni* and *P. maritimus,* which have tail filaments approximately one-third of the body length. Releases of up to two per square meter of planted area or five per infested plant are usually adequate if started in late spring, with a single repeat some 3–4 weeks later. The parasitoids will remain active up to the early autumn period, when cooler, dull conditions curtail their activities

and control ceases. However, under summer conditions, they are capable of providing up to 90% parasitism of low-level infestations on some plants, as long as humidity levels are not excessively low.

- *Influence of growing practices:* Commercial supplies of these parasitoids are frequently erratic in some European countries and their availability is limited to their seasonal activity, although reestablishment from one year's introduction to the next year is common. It should also be remembered that the parasitoid is principally used in conjunction with the far more polyphagous predator *Cryptolaemus montrouzieri*, which is more effective against large infestations of mealybugs (see p. 119).

METAPHYCUS HELVOLUS

These small (1.5 to 2.0 mm) wasps are parasitoids of various soft scale insects; females are orange-yellow and males dark brown (Figure 4.296). Native to South Africa, they have been introduced to almost all semi-tropical areas to control soft black scale (*Saisettia oleae*) on citrus and olive. They are also routinely introduced to interior atriums and conservatories for control of several species of soft scale. In temperate climates their season of activity is likely to be restricted to

Figure 4.296   Adult parasitoid wasp (*Metaphycus helvolus*).

the summer months, although small populations can overwinter to reappear the following year. Adults are relatively long-lived at 2.5–3 months, as long as an adequate food source of honeydew and young scale insects is available for host feeding.

This species is commercially available in many countries, but a license for release is required in some.

- *Life cycle:* Each female lays an average of 400 eggs at up to five per day, usually in late second or third instar nymphs. The eggs hatch after a couple of days into minute larvae that develop singly within the scale insect's body, turning it a darker brown to black color as the parasitoid matures. After 11 to 12 days at 30°C (86°F), the adult emerges through a small hole cut through the top of the scale (Figure 4.297). Female wasps are produced in larger hosts and males in smaller ones.

- *Crop/pest associations:* Control of brown soft scale (*Coccus hesperidum*) may be less effective by *M. helvolus* due to encapsulation of eggs and young larvae within the host body. Temperatures above 30°C (86°F) also tend to increase the level of encapsulation. However, at higher temperatures there is a marked increase in searching efficiency, egg

production, and host feeding that can account for greater pest mortality than by parasitism alone.

- *Influences of growing practices:* Under normal growing conditions on short season open field crops, scale insects are rarely a major pest problem. However, where plants remain *in situ* from year to year, such as citrus and olive groves, botanic gardens, conservatories, and interior landscapes, scale insects can become a serious problem. These parasitoids are relatively delicate and their level of control may be disrupted by chemical residues and other factors, including raised leaf temperatures caused by dust on the leaves. Pesticide sprays, usually targeted at other pests, can interfere with *Metaphycus* spp., causing an increase in scale insects, which in turn can require more sprays to correct the imbalance between pest and natural enemy.

Family: Trichogrammatidae

TRICHOGRAMMA SPP.

There are many species of Trichogrammatid parasitoids found throughout the world. These minute wasps (less than 1 mm in length) (Figure 4.298) feed on and parasitize the eggs of Lepidoptera and occasionally Coleoptera and Hymenoptera. The first mass

Figure 4.297   Parasitized hemispherical scale insects (*Saisettia coffeae*) showing exit hole after emergence of adult *Metaphycus helvolus*.

Figure 4.298   Adult parasitoid wasp (*Trichogramma* sp.) ready to oviposit in grain moth (*Sitotroga cerealella*) egg.

production of a *Trichogramma* species was in 1926, when 200,000 adult parasitoids were released in walnut orchards against codling moth in Ventura County, Southern California. However, most research and practical use of these parasitoids is in China, North American forest systems, Russia, and Western Europe, where they are routinely used on forestry and arable systems. This makes them the most widely used of all biological control agents in the world.

This species is commercially available in many countries, but a license for its release is required in some.

- *Life cycle:* Adult parasitoids locate moth eggs by the kairomones produced and left by their hosts. Once a suitable egg has been found, the wasp drills into it with her ovipositor and lays one or more eggs, depending on the size of the host egg. The larva has three instars and pupates inside the host egg, turning it black and killing it (Figure 4.299). Development of *Trichogramma* spp. ceases below 10°C (50°F) and the parasitoid can diapause within the host egg.
- *Crop/pest associations:* Commercial production is commonly done using the eggs of corn moths, *Ephestia kuehniella*, and *Sitotroga cerealella*. Moth eggs used

Figure 4.299   Adult parasitoid wasp (*Trichogramma* sp.) with parasitized (black) grain moth (*Sitotroga cerealella*) egg.

for mass rearing are killed by ultraviolet radiation or freezing before parasitism to ensure that no pest species are introduced. A useful semiartificial diet has also been developed in China.

- *Influence of growing practices:* *Trichogramma* spp. generally shows poor ability to disperse from release sites. To overcome this problem, very high numbers are used (75,000/hectare/week for protected crops and up to several million per hectare seasonally for field crops), and repeated releases are used through the risk periods. Parasitized host eggs may be glued to small cards for distribution by hand for ground crops or distributed by plane or helicopter for large field crops, orchards, and forests. Supplementary treatments with selective insecticides or the bacterium *Bacillus thuringiensis* may be necessary for control of any surviving caterpillars not killed as moth eggs.

Family: Crabronidae (Predatory Wasps)
CERCERIS ARENARIA (DIGGER WASPS)

Adult *Cerceris arenaria* are 9–16 mm in length and known as the sand-tailed digger wasp because they tend to nest in sandy areas where they excavate a bottle-shaped hole to rear their young. The females do this by pushing the material out of the tunnel backward using a specially flattened pygidium at the end of the abdomen. Known as solitary wasps, they frequently reuse an old nest and have been observed raiding and stealing food from burrows of their own species. Digger wasps may be distinguished from garden wasps (also known as social wasps) by their smaller size and the way they lay their wings flat over the body; garden wasps fold their wings lengthwise, often showing the abdomen. These relatively large solitary wasps prey on beetles, particularly vine weevil (*Otiorhynchus sulcatus*), which they paralyze before taking them back to the nest hole. The paralyzed prey remains alive for some time while the developing larva feeds on it.

- *Life cycle: C. arenaria* are on the wing from late spring to midsummer; females dig a tunnel in sandy soil some 30–40 cm deep, constructing a few small cells in the side walls. Each of the smaller cells is stocked with 5–15 paralyzed weevils as food for the developing larvae. A single egg is laid in each of the prepared cells; each hatches to a larva that immediately starts to feed on its supply of paralyzed prey. There is usually one generation per year in northern Europe and multiple generations further south.
- *Crop/pest associations:* There are many species of digger wasps (Figures 4.300 and 4.301) and in general they provide useful additional control of several species of weevil and some leaf-feeding beetles. Other species collect aphids, caterpillar, leafhopper, spiders, etc., to supply food in their nests. These wasps are prone to attack by several nest parasites, including some cleptoparasitic cuckoo wasps, whose developing larvae feed their young on the food collected by the digger wasps.
- *Influence of growing practices:* Intensively cultivated land is not usually

Figure 4.301 Adult digger wasp (*Cerceris sp.*). (Photograph courtesy of Dr. Ian Bedford.)

suitable for these ground-dwelling predators, but they may be found in uncultivated headlands, heath lands, and less frequented parts of amenity grounds (golf courses, parks, gardens, etc.). Although they prefer dry, firm, sandy conditions, they are equally able to make use of similar places in rural areas, where they may be found nesting in bare compacted soil in sunny situations.

Family: Vespidae (Predatory Wasp)
VESPULA VULGARIS (COMMON WASP)
The common English wasp or yellowjacket wasp (United States) is a social insect with bold black and yellow bands across its abdomen (Figure 4.302). Antennae are thick and just longer than the thorax. They usually form large colonies below ground, but occasionally nests may be made in dense bushes, hollow trees, roof spaces, and wall cavities. It is similar in appearance to the related European wasp (*Vespula germanica*), which has black dots in the yellow bands and

Figure 4.300 Adult digger wasp (*Cerceris* sp.). (Photograph courtesy of Dr. Ian Bedford.)

Figure 4.302   Adult predatory common European wasp (*Vespula vulgaris*) feeding on Tortrix moth caterpillar.

Figure 4.304   Common European wasp (*Vespula vulgaris*) queen on papier mâché nest.

Figure 4.303   Head of common European wasp (*Vespula vulgaris*) drinking juice from ripe apple.

Figure 4.305   Nest of common European wasp (*Vespula vulgaris*) made of chewed wood fibers in a house roof space.

three on the face; *V. vulgaris* has none on its face (Figure 4.303). Adult workers range from 12 to 17 mm in length and queens just over 20 mm in length. Adults feed on high-energy, sugary carbohydrates. They are also excellent predators in the spring and early summer, when they take vast numbers of insects back to the nest for their developing young.

- *Life cycle:* A nest is made of chewed wood fibers mixed with saliva to form a scalloped outer casing (Figure 4.304), which if space permits is shaped like an egg with a tapered bottom, through which the wasps enter and exit. However, nest shape may be variable depending

on available space and age of the colony (Figure 4.305). The nest contains several plates with numerous hexagonal cells, each 5–6 mm across. Each cell houses a single egg or developing larva/pupa. A large nest may hold up to 10,000 wasps by late autumn, at which time a few eggs develop into males and new queens that leave the nest and mate. Mated queens seek suitable sites in which to hibernate until spring, when they lay 20–30 eggs that develop into nonreproductive female workers who tend the new nest. The old colony dies during most winters. In warmer climates such as parts of Australia, southern Europe, the southern states of America, etc., some colonies may overwinter.

ORDER: HYMENOPTERA (PARASITOID WASPS)

- *Crop/pest associations:* Adult worker wasps contribute to pollination and control of a wide range of insect pests such as aphids and caterpillars that are fed to developing larvae along with carrion and fruit. Adult wasps feed on sweet substances such as insect honeydew, plant nectar, fruit, and pollen. They also feed on honeydew exuded by the developing larvae. The sting is a modified ovipositor and, unlike that of a bee, it has no barbs and can be used several times.

- *Influence of growing practices:* Nest sites are chosen for solitude and are rarely found in intensively cultivated land; instead, sheltered hedges, woodland, forests, and buildings are preferred. *Vespula* species also have a serious negative impact on various growing practices as they feed directly on soft fruit such as raspberries and strawberries. They occasionally attack bee hives, killing foraging worker bees and their larvae and pupae, while feeding on honey.

Beneficial pathogens

INTRODUCTION

Pathogens of insects, mites, and plants play an important role as biocontrol agents of many major pests and diseases throughout the world. Sometimes they affect a few individual hosts, but under favorable conditions, whole colonies or populations are affected and many millions of organisms killed developing as an epizootic or epidemic infection. They are generally most effective when host populations are high, in particular where pest density allows easy spread between individuals. A warm, moist environment is particularly important for nematodes and fungal pathogens.

Organisms that cause diseases of insects are known as entomopathogens and can include viruses, bacteria, fungi, and nematodes. The term mycoparasite relates to a fungus that attacks other fungi. Several bacteria are also beneficial to growers and, although naturally occurring, they may be used as biopesticides to control various plant pests. Viruses, mainly baculoviruses, are being used more widely to control specific insect pests. Due to

their extremely narrow host range, they are finding commercial applications in both conventional and organic production systems.

Beneficial pathogens are widespread throughout the world, particularly in temperate and tropical regions, where they are generally found in low numbers. However, when used as biocontrol agents, very high numbers are applied to obtain control. Isolates collected from wild strains are evaluated for their commercial potential and may be given a unique code name or number. Many of these have been DNA mapped and patented as biopesticides and commercialized throughout the world. Due to governmental registration requirements (they are often regarded as pesticides), their use as biopesticides is restricted in some countries. However, being of natural origin, many beneficial pathogens are eligible for use in organic growing systems. They are also widely used in conventional growing systems, particularly on edible crops as they usually have no harvest interval (the time that must be allowed between application and harvest to ensure that any residues are below set limits). Representative beneficial pathogens are described here, including some nematodes, fungi, bacteria, and viruses.

FAMILY: BACULOVIRIDAE

Viruses

NUCLEOPOLYHEDROVIRUSES (NPVs) AND GRANULOVIRUSES (GVs)

These are baculoviruses that have double-stranded DNA embedded in a crystalline proteinaceous occlusion body. They are mostly found in lepidopteran larvae, although some viruses attack other insects including caddisfly, cranefly, mosquitoes, and sawflies. Historically, this group of insect-killing pathogens has been recorded for over 2,000 years, initially from infected silkworm caterpillars (*Bombix mori*). Although the actual causative organisms have been known for over 100 years, they have only recently been genetically described. They are also most likely to account for the epizootics that have frequently infected caterpillar populations in various parts of the world, killing many millions of insects.

There are two major types of baculoviruses: nucleopolyhedroviruses (NPVs) and granuloviruses (GVs). Within the NPVs are rod-shaped particles (virions) protected in a membranous envelope containing either single or groups of virions embedded in a proteinaceous, polyhedral-shaped occlusion body. They range in size from 1 to 5 nm (nanometers) in diameter and form in the nuclei of infected cells (1,000,000 nm = 1 mm). GVs are structurally similar to NPVs, but the virions are occluded individually in small bodies referred to as granules that have an average size of approximately 150 nm × 400–600 nm.

- *Mode of action:* All viruses must reproduce in the cells of living organisms by invading individual cells to replicate, firstly in the nucleus and, after rupturing the nuclear membrane, in the cytoplasm.

The predominant route of viral infection is by ingestion of inoculated plant material. Once in the alkaline juices of the caterpillar midgut, the protective granules and polyhedra surrounding the virions are rapidly digested. The released virions pass through the peritrophic membrane to infect various body cells. As a viral infection progresses, the body may become swollen and take on a glossy appearance (Figure 5.1). With a light microscope, it may be possible to see white polyhedra within swollen cells. The final stages of the disease are seen as a liquefaction of the caterpillar body, which typically hangs head down by the prolegs or in an inverted "V" (Figure 5.2). The cuticle then ruptures after a few hours, releasing many millions of virus particles. There are two types of GVs, referred to as fast and slow. Caterpillars infected with slow GVs die at the final larval instar stage, regardless of when they become infected. Infected larvae appear as having creamy white to yellow patches of fat cells packed with viral granules. Fast GVs kill at the same rate as NPVs.

- *Crop/pest associations:* Entomopathogenic viruses often have a very narrow insect host range, usually restricted to just one or two species, and none are known to have any vertebrate hosts. The occlusion bodies offer great

Figure 5.1   Small cabbage white butterfly caterpillars (*Pieris rapae*); the pale one is infected with *P. rapae* granulovirus (*PrGv*).

Figure 5.3   Codling moth (*Cydia pomonella*) killed by *C. pomonella* granulovirus (*Cp*Gv).

Figure 5.2   Small cabbage white butterfly caterpillar (*Pieris rapae*) showing late-stage infection with *Pr*Gv; infected insects typically hang head down.

Figure 5.4   Codling moth (*Cydia pomonella*) killed by *Cp*Gv; note pips (seeds) in apple core.

protection for the virus. Many moth species have only one or two generations each year, so the virus must be able to replicate and then be able to survive until the following year. Some viruses are known to have survived for several decades in the soil before entering a suitable host insect and still be infectious.

- *Influence of growing practices:* Baculoviruses are generally regarded as specific insect pathogens that show a high level of environmental safety and pest selectivity. They have great potential as bioinsecticides that can be targeted to a particular pest. Commercial formulations are known by the name of the target organism followed by either GV or NPV. For example, in Brazil over 1 million ha of soybeans are treated annually to control the velvet-bean caterpillar (*Anticarsia gemmatalis*) with an *Ag*NPV baculovirus. In several European countries virus preparations

are commercially available to control apple codling moth larvae (*Cydia pomonella*) with a *Cp*GV (Figures 5.3 and 5.4), corn earworm (*Helicoverpa zea*) with an *Hz*NPV, and European pine sawfly (*Neodiprion sertifer*) with an *Ns*NPV. It is likely that in the future many more baculoviruses will be registered for commercial use as bioinsecticides throughout the world.

Bacteria

Bacteria are simple unicellular organisms, lacking a cell nucleus and mitochondria. Reproduction is by binary division which can occur anywhere suitable for bacterial growth. They are typically just a few microns (μ) in length (1000 μm = 1 mm) and have a range of shapes including rods, spheres, and spirals. The majority of entomopathogenic bacteria are in the family Bacillaceae and belong to the genus *Bacillus*. Within the *Bacillus* species there are numerous subspecies and strains, each with a different range of activity against widely differing target pests. Although naturally occurring, several isolates are in commercial production where a liquid fermentation type process is used. The resulting bacteria can be formulated as liquids or granular powders for application by conventional spraying. In the majority of cases, these commercial products have an active storage life of 1–2 years before they lose significant levels of activity.

Figure 5.5   **Gray mold (*Botrytis cinerea*) on ripe raspberry fruit.**

FAMILY: BACILLACEAE

Bacillus subtilis

Bacillus subtilis is a ubiquitous, saprophytic, Gram-positive, rod-shaped, aerobic bacterium that is commonly recovered from soil and decomposing plant material. It was first characterized in 1835. Different strains are commercially produced as biocontrol agents against a range of plant pathogens, while others are used in industrial processes (i.e., in the production of biological detergents) and also for the treatment of water to minimize or control harmful pathogens. The biocontrol strains form a distinct group referred to as the variant or subspecies *B. subtilis* var. *amyloliquefaciens*. There are three general categories of biocontrol products: those that are applied to the foliage of plants to control and suppress various plant pathogens such as *Botrytis* (Figures 5.5–5.8), downy and powdery mildews, rusts, etc.; those applied to the soil or transplant mix when seeding to control or suppress root and soil pathogens such as *Fusarium, Pythium, Rhizoctonia,* and *Sclerotinia* spp.; and those applied directly as seed treatments for protection from soil pathogens such as *Fusarium* and *Rhizoctonia*.

- *Life cycle:* In the vegetative stage the bacterium uses flagellae that enable it to have some localized movement. In adverse conditions it is able to produce endospores that are very resistant to

Figure 5.6   **Gray mold (*Botrytis cinerea*) mycelium on damaged pea pod.**

extremes of light and temperature. These spores enable the organism to persist for long periods in the environment until the return of favorable conditions. The control of foliar pathogens is by the production of lipopeptides, which lyse (destroy) spore membranes, causing cell death of the fungal pathogen (Figures 5.9 and 5.10). Other metabolites stop protein and cell membrane synthesis

Figure 5.7 Young tomato fruit with ghost spotting caused by *Botrytis cinerea;* note fungal mycelium on remains of old flower and infection to bruised lower portion of fruit.

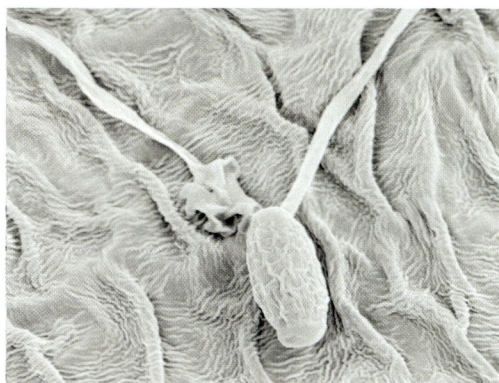

Figure 5.9 *Botrytis cinerea* fungal spore showing germinating hyphae on leaf surface. (© 2013 Bayer CropScience LP. Used with permission.)

Figure 5.8 Rose flowers with *Botrytis cinerea* ghost spotting following damage caused by hail storm.

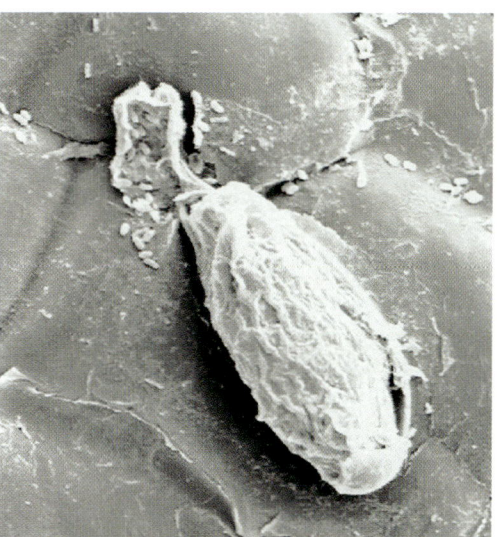

Figure 5.10 *Botrytis cinerea* fungal spore infected with beneficial bacterial pathogen *Bacillus subtilis* showing cell wall destroyed by enzyme lipopeptide action. (© 2013 Bayer CropScience LP. Used with permission.)

of bacterial pathogens. Below ground, other than the modes of action described before, the bacteria colonize developing root systems by feeding on root exudates and in this way compete with plant pathogens.

- *Crop/pest associations:* The novel and multiple site mode of action of *B. subtilis* makes it difficult for fungal pathogens to develop resistance. In addition, the *B. subtilis* var. *amyloliquefaciens* strains produce several elicitor compounds (jasmonic acid and/or ethylene) that activate plant defenses through the process known as "induced systemic resistance" (ISR), thus giving better protection against

FAMILY: BACILLACEAE

several pathogens and faster recovery following disease development. High levels of nutrient organic material in soil and on plants encourage growth and survival of the bacteria.

- *Influences of growing practices:* B. subtilis is compatible with most insecticides and fungicides such as copper, sulfur, and foliar-applied micronutrients. Because it is nontoxic to beneficial insects, mites, bees, and other nontarget organisms, it is suitable for use in IPM (integrated pest management) programs. Most regulatory authorities have approved one or more *B. subtilis* products for use on a wide range of crops, including organic production. None of the commercial products are known human pathogens or disease-causing agents. However, it has been reported to cause dermal allergic reactions or hypersensitivity in some individuals.

Bacillus thuringiensis

Bacillus thuringiensis (*Bt*) is a spore-forming bacterium that was first isolated from infected silkworms in Japan in 1902. In its vegetative state, this Gram-positive aerobic bacterium measures 3–5 µm × 1.0–1.2 µm. Several subspecies vary in their activity against certain insects such as caterpillars, mosquito larvae, and certain beetle larvae including their adults. This range of activity is determined by the matrix of complex protein toxin crystals within the bacterial cell. Bipyramidal crystals (like two Egyptian pyramids fixed base to base) range in size and contain up to five different toxic protein subunits. However, although no single strain has activity against all the target pests, it is now possible to combine two strains together and widen the range of activity. This process is known as transconjugation and has led to some patented bioinsecticide products. Through genetic manipulation it is also possible to transfer the toxin-producing protein genes into plants, thus protecting them from caterpillar attack.

- *Mode of action:* The active ingredients of formulated *Bt* are viable cells and toxic protein crystals. Activity within a larval insect is dependent on ingestion of the crystals. Once ingested, pH conditions (pH > 9) and gut enzymes of the insect rapidly break the crystals down to toxic subunits that attack the midgut. Paralysis of the gut occurs within a few hours. Feeding and movement cease soon after, and cells of the gut become extensively damaged, leading to insect death after a day or so (Figures 5.11 and 5.12). Dead insects fall to the ground or stick to leaves and slowly disintegrate (Figure 5.13).

Figure 5.11   **Top image: healthy small cabbage white butterfly (*Pieris rapae*); bottom: the same caterpillar a few days later following infection by beneficial pathogen *Bacillus thuringiensis*.**

Figure 5.12   **A tomato moth (*Lacanobia oleracea*) infected with *Bacillus thuringiensis*.**

Figure 5.13   Cabbage white butterfly caterpillar (*Pieris brassicae*) 5 days after treatment with *Bacillus thuringiensis*.

- *Crop/pest associations:* Unprotected, leaf-feeding caterpillars are the easiest target to control and are the most common targets of *Bt* worldwide. Insects sheltering within folded or rolled leaves, plant stems, or in silken webs are more difficult to reach. Similarly, insects that have burrowed into leaves or stems are impossible to control with sprays of *Bt*. Because of the specific requirements of the protein crystals, they are only toxic to specific insects. *Bt* is much safer to nontarget organisms (including man) than many pesticides.
- *Influence of growing practices:* Although *Bt* is usually applied as high-volume sprays in order to wet the foliage thoroughly, treatments of thermal fogs and other ultra low-volume methods have given excellent results. As the bacteria must be ingested for activity,

many growers have added an apetant in the form of unrefined sugar to the spray, which encourages caterpillars to feed actively on the treated surface. The toxic crystals may be denatured by ultraviolet (UV) radiation, which may occur in field crops, orchards, and in forestry after a few days of sunshine. Under glass, the activity usually remains for up to a week after application.

Fungi

When fungi infect insects and mites, they are known as entomopathogens. All entomopathogenic fungi require high humidity for initial spore germination and penetration. They are one of the most common causes of disease found in nature, particularly in warm, humid environments. The germ tube of a fungal spore can penetrate an insect cuticle and colonize the host tissues. Fungal hyphae grow throughout the body, leading to host death after a few days. In favorable conditions, more fungal spores will be formed either internally or externally. Many hundreds of thousands of spores may be produced from each infected host; under the right conditions, they can spread to other insects as an epizootic infection.

Several fungal pathogens are commercially produced as bioinsecticides and, although naturally occurring throughout the world, often require registration similar to chemical pesticides. This process may hinder their availability. Fungi that infect other fungi are known as mycoparasites. Commercially, they are used mainly to protect against a range of fungal plant pathogens and as part of an IPM program to reduce the reliance on conventional chemical fungicides. There is increasing evidence that many entomopathogenic fungi also provide additional protection to plants from plant pathogenic fungi. A possible mechanism for this is the (beneficial) fungus outcompeting the plant pathogenic fungi for space and nutrients. Insects attracted to roots to feed on are inevitably infected and killed, thus providing a useful source of nitrogen and other nutrients. Similarly, foliar-applied entomopathogenic fungi aimed at insects or mites might also provide some protection against a range of fungi that cause leaf, stem, and flower diseases.

FAMILY: HYPOCREACEAE

Beauveria bassiana

There are several naturally occurring species and strains of insect pathogens within the genus *Beauveria*. *B. bassiana* is in commercial production and is now a registered product in several countries throughout the world. This fungus is used as a contact mycoinsecticide but survives a relatively short period of time when exposed on a leaf surface. Most commercial formulations are based on conidiospores harvested from a solid agar substrate. These may be packaged as an emulsifiable suspension in oil to extend spore survival on leaves or as a dry dispersible powder that is made up in water.

Figure 5.14   Soft brown scale insects (*Coccus hesperidum*) infected with beneficial fungal pathogen *Beauveria bassiana*.

Figure 5.15   Cottony cushion scale insects (*Icerya purchase*) infected with *Beauveria bassiana*.

- *Life cycle:* The conidiospore is nonmotile and asexual. Once contact has been made with the target organism, the spore adheres to the cuticle where it germinates to produce a germ tube that penetrates the host's body. Blastospores are formed by budding of conidiophores in the hemocoel. These germinate to initiate secondary infections leading to rapid death of the host organism due to production of various mycotoxins and loss of body fluids. This process can kill the insect in 3–7 days, leaving a mass of spores that can spread to other insects (Figures 5.14 and 5.15).
- *Crop/pest associations:* B. bassiana has a very wide host range and is reported to control aphids, leaf hoppers, mealybugs, scale insects, sciarid flies, scatella flies, thrips, whiteflies, and vine weevil larvae. Under ideal conditions it will also infect scale insects, spider mites, and leaf-feeding caterpillars. *Beauveri* spp. have

low mammalian toxicity and can infect a high proportion of pest organisms, often resulting in fungal epizootics. Generally, *B. bassiana* integrates well in IPM systems and is safe to most beneficial organisms; however, *Aphidoletes aphidimyza* larval stages are highly susceptible. Bumblebees are not affected and trials have shown that bees can be used to disseminate various biofungicides to control a range of insect and mite pests living on plants and particularly those associated with flowers such as thrips.

- *Influence on growing practices: B. bassiana* can be formulated to infect insects over a range of conditions. Infected insects, like healthy ones, may cause a cosmetic problem on certain crops due to the visible presence of their bodies. Fungicides used to control plant pathogens may disrupt these mycoinsecticides; however, several insecticides and, in particular, chitin inhibiting insect growth regulators are compatible and can be safely used within an IPM program.

Gliocladium catenulatum, G. virens
Various species of *Gliocladium* are commonly found as saprophytes in many environments, but they are also pathogenic on several fungi including the cultivated mushroom (*Agaricus bisporus*). *Gliocladium* has been isolated from organic matter and plant debris from around the world; a few species are extremely potent biological control agents against a range of plant pathogens. Isolates from at least two species have been registered as commercial biofungicides: *G. catenulatum* strain J1446 and *G. virens* strain GL-21. *Gliocladium* produces transparent, slimy septate hyphae. Conidiophores are erect and branch repeatedly at their apices. The terminal branches give rise to bottle-shaped projections called phialides. Conidia or phialospores are one celled, ovoid to cylindrical, and produced in a terminal head in a single slimy ball or occasionally in a loose column. Spores are dispersed in water or carried with soil and organic matter and by soil-dwelling insects and mites.

This species is commercially available in many countries, but a license for its use is required in some.

- *Life cycle:* *Gliocladium* produces quick growing, fine, hair-like hyphae from germinating spores, which may contact a plant pathogen. However, commercial formulations can contain up to 2×108 cfu/g (colony forming units per gram), greatly increasing the chances of this fungus contacting suitable hosts. The hyphae can kill the host fungus by enzyme activity but without penetration. However, haustoria have been observed on the hyphal tips of *Rhizoctonia solani*, although most activity has been seen as

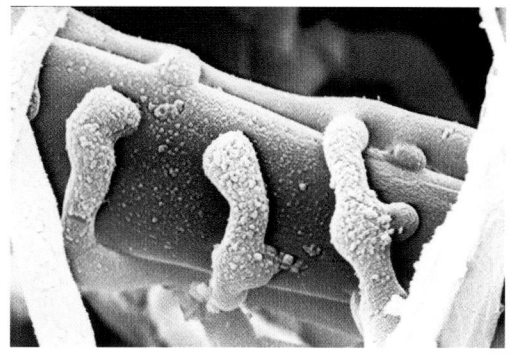

Figure 5.16 *Gliocladium catenulatum* wrapping around a plant root in a beneficial association to protect against fungal attack by *Rhizoctonia solani*. (Electron micrograph courtesy of Verdera Oy, Finland.)

coiling of *Gliocladium* hyphae around the host hyphae (Figure 5.16). There is also evidence that *Gliocladium* can form a close relationship with healthy roots, giving protection against disease attack. Colonization of the root zone seems to be a particularly important feature of the ecology of *Gliocladium*. However, a truly symbiotic relationship has not been demonstrated with any of its associated organisms. *Gliocladium* is active at temperatures between 5°C and 34°C with best activity at 15°C to 25°C. Temperatures above 42°C can kill the fungus, spores, and mycelium.

- *Crop/pest associations:* Commercial *Gliocladium* strains can be used as biofungicides in a wide range of situations, including the control and prevention of diseases on amenity turf, herbs, ornamentals, tree and shrub seedlings, vegetables, etc. The main target plant pathogens are those that cause damping-off diseases (Figure 5.17) and also some that cause foliar and stem diseases (Figure 5.18). Pathogens involved are in the genera *Alternaria, Botrytis, Didymella, (Mycosphaerella), Fusarium, Pythium, Phytophthora, Rhizoctonia,* and *Sclerotinia*. *Gliocladium* spores can survive for a time on treated plant surfaces, but repeated applications are

Figure 5.17   Typical symptoms of a damping off disease. In this instance *Pythium* infection on maize seedling showing significant loss of root and wilting growth.

Figure 5.18   Gray mold (*Botrytis cinerea*) infecting tomato stem; this is a secondary infection on physically damaged tissue.

required for continued control/protection. Disease prevention and suppression on plants occurs when the biofungicide outcompetes the disease pathogen for nutrients and, once established, excludes other pathogens invading the root area. Biofungicidal activity occurs through a combination of several mechanisms. These include hyperparasitism or mycoparasitism (the fungus attacking the plant pathogen) and antibiosis. The production of antibiotic gliotoxin has been shown to occur with some strains of *Gliocladium*. However, studies with the commercial strain J1446 have indicated a complete absence of antibiosis, which simplifies its use and approval as it is unlikely to be significant when human or veterinary medicines are considered. *Gliocladium* has not been reported as the causative agent of any disease in man or animals.

- *Influence of growing practices:* Commercial products containing fungal spores and mycelium are supplied in granular or powder form that can be incorporated with soil, compost, or other growing media prior to seeding, planting, or transplanting. It can also be applied as a drench, irrigation treatment, or spray to existing plants. Because it is a fungal pathogen, care should be taken when using other fungicidal preparations; the manufacturers and suppliers of commercial *Gliocladium* products should have up-to-date information on the use of their product. As a general guide, where *Gliocladium* is used within the rhizosphere (in close association with the roots), it is unlikely to be affected by foliar fungicides, but those applied to the soil might adversely affect it. However, if it is used as a foliar treatment against *Botrytis* or *Didymella* (*Mycosphaerella*), a period of up to 4 days should be allowed between treatments of a biofungicide and conventional (chemical) fungicide.

Trichoderma asperellum

Trichoderma asperellum is a filamentous, asexual fungus, originally called *T. viride* (some isolates of *T. harzianum* were similarly classified as *T. viride*). Further taxonomic studies have identified a new species: *T. asperelloides,* which is very closely related to *T. asperellum*. These fungi are commonly isolated from soil, decomposing, and healthy plant matter throughout temperate and tropical areas. Their antifungal abilities have been known since the 1930s and they are now commercially available to suppress and control a range of plant diseases, including those caused by various soil-borne pathogens in the genera *Pythium, Phytophthora, Rhizoctonia, Sclerotinia, Thielaviopsis,* and *Verticillium*.

Trichoderma spp. are powerful antagonists of other fungi and are able to parasitize (mycoparasitism) and inhibit their growth and development (Figures 5.19 and 5.20). These beneficial fungi can colonize soil and artificial growing media, establishing around plant roots to compete with plant pathogenic fungi for space and nutrients. Moreover, *T.*

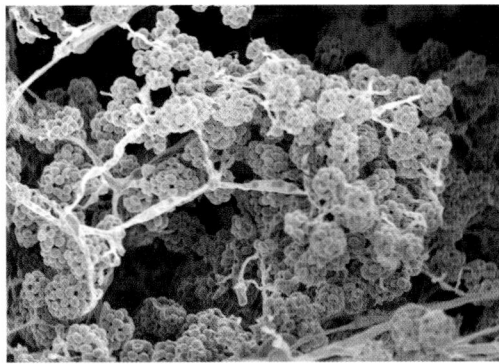

Figure 5.20   Mycelium growth with clusters of *Trichoderma asperellum* spores. Each spore is an infective body or colony-forming unit. (Scanning electron micrograph courtesy Lourdes Cotxarrera, University of Barcelona, Spain.)

asperellum also uses enzymes to attack the cell walls of many pathogenic fungi and enhance the host plant's systemic resistance.

- *Life cycle:* *T. asperellum* forms an association with plant roots where it competes with other soil microorganisms for nutrients and space. It also inhibits and degrades several enzymes that are produced by plant pathogenic fungi used to penetrate healthy roots, conferring useful protection against invading plant diseases. The fungi grow toward hyphae of other fungi, coil around them, and degrade the cell walls of the target pathogen by secretion of various hydrolytic enzymes. The weakened cell wall allows *Trichoderma* to extract nutrients for its own growth and development. Antibiotics that can kill or severely inhibit some plant pathogens may be produced by some strains of *Trichoderma*.
- *Crop/pest associations:* *Trichoderma* hyphae grow on plant roots, building a physical barrier against invasive pathogens. Some strains also penetrate the root epidermis and outer cortex of roots, which triggers the whole plant's immune system. This is known as induced systemic resistance (ISR) or systemic acquired resistance (SAR) and

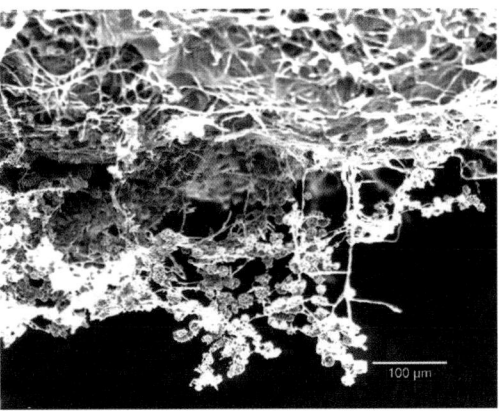

Figure 5.19   *Trichoderma asperellum* colonizing sclerotia of *Sclerotinia sclerotiorum*—showing degradation of the sclerotia with resultant profuse growth and sporulation of the beneficial fungus. (Scanning electron micrograph courtesy Isabel Trillas, Biocontrol Technologies, S. L. Barcelona, Spain.)

can imply protection throughout the plant against a range of foliar pathogens. Analysis of the signaling mechanism indicates the involvement of salicylic acid, jasmonic acid, and ethylene as the most important active molecules.

- *Influence of growing practices:* Root colonization by *Trichoderma* spp. frequently enhances root growth and development, leading to increased crop productivity by activating the host's resistance to abiotic stress and helping the uptake of nutrients. When applied in the early stages of plant growth, including field transplants, the use of chemical fungicides is greatly reduced. This has an added advantage of reducing fungicide resistance pressure. Continuing trials have shown good compatibility with fertilizers, chemical insecticides, beneficial nematodes, predatory mites, and beetles. Certain fungicides, such as those with activity against ascomycetes and basidiomycetes, can be harmful to *Trichoderma,* so an interval of 10 to 14 days should be allowed between applications. When appropriate, mycorrhizae should be applied first, allowing 2 to 4 weeks before applying *Trichoderma.*

FAMILY: CLAVICIPITACEAE

Lecanicillium muscarium, Lecanicillium longisporum (syn. *Verticillium lecanii*)
The fungus genus *Lecanicillium* contains several insect pathogenic species that differ in their host range and may be very host specific. Its fluffy white halo-like appearance can be found around affected insects (Figure 5.21). It is common throughout the tropics, where it may be found infecting aphids, mites, scale insects, and whitefly. It is also a saprophyte and hyperparasitoid of leaf spot pathogens, powdery mildews, and rusts. However, it cannot parasitize plants, birds, fish, or mammals.

Figure 5.21   Alate aphid killed by *Lecanicillium lecanii*.

This species is commercially available in many countries but a license for its use is required in some.

- *Life cycle:* Spores from a commercial product or from a parasitized insect must be in direct contact with the host cuticle. Spores are sticky, so they will adhere to a passing insect. Germination and subsequent penetration of the host cuticle requires a high humidity (above 95%) for about 12 hours. Once the fungus has begun growing in the insect, a high humidity is no longer required. Infected aphids continue feeding and can even produce healthy live young until quite late in the parasitism process. After 5–10 days, depending on temperature and humidity, the insect dies and is covered by a fluffy white mass of spores. For further spread a high humidity is again required (Figure 5.22).
- *Crop/pest associations:* Commercially, there are two strains of *Lecanicillium* available: *L. longisporum,* which is more specific for aphids, and *L. muscarium,* which can infect both thrips and whitefly (Figure 5.23). If several pest species are present on the crop, both strains can be mixed to provide a wider spectrum of activity. They can also be mixed with *Bacillus thuringiensis* for caterpillar control.

Figure 5.22   Various stages of infection by entomopathogenc fungus (*Lecanicillium lecanii*); images were taken 1 day apart.

Figure 5.23   Glasshouse whitefly (*Trialeurodes vaporariorum*) infected with *Lecanicillium muscorum*.

- *Influence of growing practices:* Temperature is not critical between about 15°C and 28°C (59°F and 82°F). Temperatures above (up to 38°C [100°F]) and below (down to 2°C [36°F]) this range will not kill the fungus but will arrest its development until suitable conditions return. Humidity, however, is the most important factor influencing germination and infection of *Lecanicillium* spp. in a pest population and, because of this, the spread of the pathogen may be limited by the relative humidity. In cucumber crops favorable conditions can be achieved most nights, but in tomato and most other crops the humidity requirement may only be suitable during spring and autumn, especially during the evening and early morning. Research has shown that

increasing humidity by fogging water above the crop for 40 hours per week will provide the necessary conditions without causing undue problems to the plants or increasing the incidence of plant diseases.

Metarhizium anisopliae

Metarhizium anisopliae, commonly known as green muscardine fungus, was first isolated in Russia in 1869 from *Anisoplia austriaca*, the wheat chafer or grain beetle. It is a soil-inhabiting entomopathogen found naturally throughout the world and is in commercial production as a biopesticide (mycoinsecticide). It has activity against a range of economically damaging pests, including the larvae of various root-feeding beetles such as vine weevil (Figure 5.24), the larvae/pupae of soil-pupating insects such as thrips, adults and nymphs of locusts, termites (Figures 5.25 and 5.26), etc.

Figure 5.24   Vine weevil larvae (*Otiorhynchus sulcatus*) infected with *Metarhizium anisopliae*.

Figure 5.25   Termites (*Reticulitermes flavipes*) infected with entomopathogenc fungus (*Metarhizium anisopliae*).

Figure 5.26   Close-up of termites (*Reticulitermes flavipes*) infected with *Metarhizium anisopliae*.

This species is commercially available in many countries, but a license for use is required in some.

- *Life cycle:* Once spores of the fungus make contact with a host insect, they germinate, producing thread-like hyphae that penetrate the cuticle, usually through breathing spiracles and between plates in the exoskeleton. Further mycelial growth occurs within the host body, eventually killing the insect after a few days. This lethal effect is enhanced by the production of secondary metabolites called destruxins (cyclic peptides) that turn the insect body a reddish color. As long as the humidity is high enough, a white fungal mold develops on the dead host and turns it green as spores are produced a few days later.

- *Crop/pest associations:* There are many isolates or strains of *M. anisopliae*, each with different activity against a range of insect pests. Recent trials in the UK have shown an improvement in control of vine weevil larvae when mixed with entomopathogenic nematodes, often extending the period of control/protection to several months. This fungus is considered safe to invertebrates and is being evaluated for the control of insects that can affect animal and human health, such as control of flies in animal husbandry and the treatment of mosquito breeding sites to control malaria.

- *Influence of growing practices:* Once inside the host body, humidity is not critical. By formulating *M. anisopliae* in oil-based carriers, the fungus can be applied as minute droplets by airplane, giving excellent control of locust and grasshoppers in Africa and Australia. Other formulations include rice grains on which the fungus is grown that can be incorporated in soil or growing media. Wettable powder products can be sprayed in water directly to plants for control of various leaf-feeding pests. Compatibility with most insecticides is generally good and can be used as a method of enhancing control, although caution should be taken with fungicide sprays. Ideally, 4–7 days should be allowed between application of the fungus and a fungicide.

FAMILY: PHAEOSPHAERIACEAE

Ampelomyces quisqualis

This naturally occurring fungus is a parasite of a range of powdery mildew diseases (Figures 5.27–5.30). (Powdery mildews are common and important diseases that affect a wide range of plants.) The hyphae and conidiophores of the pathogen swell to several times their normal size and take on a dull, amber-gray color (Figures 5.31 and 5.32). The pycnidia develop within the host fungus.

Figure 5.27   American powdery mildew (*Sphaerotheca mors-uvae*) on gooseberry fruit.

Figure 5.28 Powdery mildew (*Sphaerotheca fuliginea*) on cucumber leaves.

Figure 5.29 Wheat powdery mildew (*Blumeria graminis* f. sp. *tritici* syn. *Erysiphe graminis* f. sp. *tritici*) with black chasmothecia (cleistothecia).

Figure 5.30 Close-up of powdery mildew (*Erysiphe graminis*) on barley leaf.

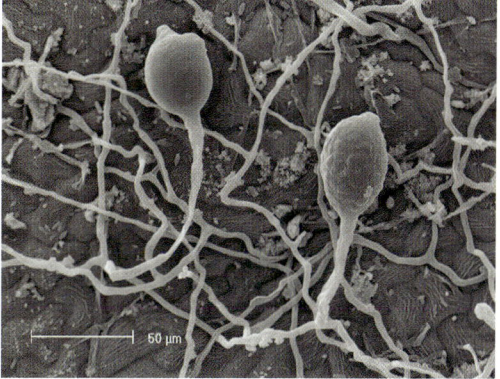

Figure 5.31 Powdery mildew conidi-ophore infected with *Ampelomyces quisqualis*. (Photograph courtesy of CBC Europe spa.)

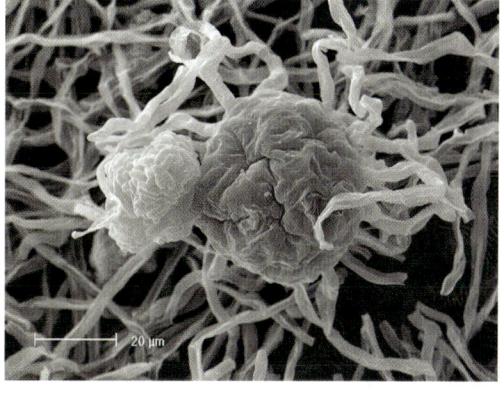

Figure 5.32 Powdery mildew chasmothecia par-asitized by *Ampelomyces quisqualis*. (Photograph courtesy of CBC Europe spa.)

Isolate M-10 was collected in the 1930s from powdery mildew-infected plants in Israel and is commercialized worldwide as AQ10.

- *Life cycle:* The pathogen requires warm (20°C to 30°C) and humid (ideally, free water) conditions to infect and penetrate

the cell walls of the powdery mildew, usually within 24 hours of initial contact. Hyphal growth occurs within the mildew for 7 to 10 days; during this time there

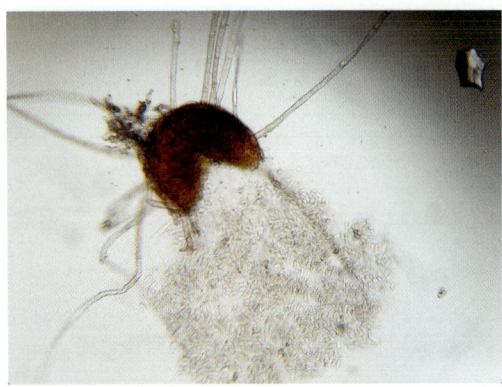

Figure 5.33 Beneficial fungal pathogen (*A. quisqualis*) coming out of ruptured powdery mildew chasmothecia. (Photograph courtesy of CBC Europe spa.)

is little exterior sign of the pathogen. Asexual fruiting bodies (pycnidia) are produced within the hyphae of the mildew. Chasmothecia (the sexual stage of the pathogen that releases spores on decay or rupture [Figure 5.33]) gives the dying mildew a flattened, creamy gray appearance. The presence of free water (crop spraying or dew) releases conidia from ripe pycnidia, which are spread by water splash to mildew colonies. It is possible for multiple generations to occur each season. *A. quisqualis* can overwinter as mature pycnidia on plant debris and inert surfaces.

- *Crop/pest associations:* Applications of *A. quisqualis* should be used as a preventive measure or at the earliest stages of mildew development for successful control. Growth of *A. quisqualis* reduces powdery mildew growth and may eventually kill the mildew colony. If the plants are severely affected, they will be slow to recover. Many organic growers use this biofungicide either alone or more commonly in rotation with sulfur. Nonorganic production (commonly known as "conventional" growing) tends to alternate chemical fungicides with biopesticides as part of an IPM program to reduce reliance and to minimize resistance issues. Due to the very short harvest interval (HI), many countries have set a zero day HI for this and other biopesticides, making them ideal to use right up the time of harvesting on edible crops like strawberries.

- *Influence of growing practices:* Although several conventional and biological fungicides can be integrated with *A. quisqualis*, care must be taken when spraying any pesticide without first confirming its compatibility. This is particularly so if sulfur is included in the spray program, as small amounts of this chemical will adversely affect the biocontrol agent. The presence of free water, for infection to occur, must be taken into account and rapidly drying conditions will greatly reduce efficacy of this mycoparasite. Good commercial results have been achieved on a range of protected crops where the small amount of powdery mildew needed for the establishment of the pathogen does not detract from the crop's appearance. Repeated applications are required for adequate mildew control.

FAMILY: ENTOMOPHTHORACEAE

Furia sciarae

There are several species of *Furia* that infect sciarid flies and other soil-associated insects. Infection of the larva causes it to change its normal behavior (which is to avoid light by living in the soil where it feeds on plant roots) and move to the soil surface, where it dies. Dead, infected larvae on the soil surface appear as milky-white threads 8–10 mm in length and up to 2 mm in diameter (Figure 5.34). There are no commercial products containing these fungi at present, but methods of encouraging their development are being investigated.

- *Life cycle:* Conidiophores develop on the surface of sciarid larva within 24 hours

Figure 5.34   Sciarid fly larvae (*Bradysia pauper*) on compost surface, infected with beneficial fungal pathogen (*Furia sciarae*).

Figure 5.35   Sciarid fly larva (*Bradysia paupera*) infected with *F. sciarae*.

of infection and these produce sticky conidia. Larvae moving through or across the soil may become infected by contact with spores. Adult flies are also likely to carry the fungus on their bodies. Conidial contact with the host larvae is followed by penetration of the cuticle and invasion of the body tissues. As the host is overcome with fungal growth, it changes to a gray/brown color and moves upward, finally to rest in the open (Figure 5.35). Here it turns milky-white as a result of fungal sporulation, and many more conidia are produced from the conidiophores that are spread by contact with other insects and water splash.

- *Crop/pest associations:* This group of soil-active entomopathogenic fungi are mainly found on sciarid larvae (*Bradysia* spp.), but it is possible that other organisms

inhabiting similar environments, such as shore flies (*Scatella stagnalis* and *S. tenuicosta*), could also be hosts. *Furia* spp. are also associated with several aphid species and may be found throughout the temperate regions, particularly on cereal crops. The occurrence of this pathogen in the UK has been associated with peat-based growing media. It is likely that sciarids in similar media, such as coir and bark, will also be affected.

- *Influence of growing practices:* Fungicides applied to control root diseases such as *Pythium*, *Phytophthora*, and *Thielaviopsis*, which may be spread by sciarid flies, could seriously affect these naturally occurring entomopathogenic fungi. Resting stages have been found for some *Furia* spp. and are likely to be produced in adverse conditions, enabling them to survive between cropping cycles. In suitable conditions a high level of infection is possible, resulting in fungal epizootics that may persist between crops. The presence of dead bodies on the soil surface may cause cosmetic problems in pot plant production, but they tend to disintegrate after a few days, leaving little sign of insect or fungal presence.

Pandora neoaphidis (syn. *Erynia neoaphidis*)
Pandora neoaphidis is one of a group of entomopathogenic fungi. This and related species are found throughout the world, where they readily develop epizootic infections (Figures 5.36 and 5.37). Indeed, in some years entomopathogenic fungi kill more insects than any of the arthropod parasitoids and predators.

- *Life cycle:* Conidia contact the host and germinate in high humidities. The germ tube penetrates the cuticle and the body tissue is invaded. Death occurs a day or so after infection, but outward signs may take several days to appear. The cadaver is attached to the plant by thin stalk-like

Figure 5.36   Colony of blackcurrant-sowthistle aphids (*Hyperomyzus lactucae*) infected with entomopathogenc fungus (*Pandora neoaphidis*).

Figure 5.37   *Hyperomyzus lactucae* infected with *Pandora neoaphidis*; note green, healthy aphids compared to rusty brown, infected aphids.

Figure 5.38   Rose aphids (*Macrosiphum rosa*) infected with *Pandora neoaphidis*.

- *Influence of growing practices:* *Erynia* and related fungi in the Entomophthoraceae are highly virulent and can provide high levels of pest control. They are most common where aphids are more abundant and usually make their appearance a few weeks after the initial aphid infestation. Overhead irrigation increases the incidence and spread of these fungi due to the raised humidity and the splash spread of the spores. The use of fungicides to control plant pathogens can adversely affect most of these fungal insect pathogens.

Nematodes

EELWORMS OR ROUNDWORMS

These often minute organisms are relatively simple—bilaterally symmetrical, elongated, and tapered at both ends. The species described here are facultative parasitoids (i.e., they can live as saprophytes as well as parasitoids). Although found in nature, they can be mass produced on artificial diets using a liquid medium fermentation type process and are commercially used as biological control agents. Unlike plant parasitoid nematodes, these entomopathogenic species have symbiotic bacteria in their alimentary tract. These produce a toxin and it is this that is the lethal agent. Once the nematode has entered the host and is feeding on its hemolymph, it defecates a small pellet containing the pathogenic bacteria, which, under the right temperature conditions, kill

rhizoids that terminate in a flattened disc structure. Ripe conidiophores rupture and forcibly discharge their conidia into the air to spread the pathogen.

- *Crop/pest associations:* Dead aphids take on a pale brown to orange-red color as they become covered in fine velvety spores (Figure 5.38). Other related fungal species may be found infecting Diptera, Homoptera, Lepidoptera, and Orthoptera. Although they are regarded as beneficial by most growers, high numbers on edible leafy salad crops may cause cosmetic damage and a reduction in marketability. To date there are no commercial preparations based on *Erynia* spp.

the host after only 2–3 days. The nematodes then reproduce in the soup of bacteria and hemolymph, leaving the cadaver as third-stage infective larvae. These are unusually resistant to adverse environmental conditions and can survive several months as "dauer larvae."

FAMILY: HETERORHABDITIDAE

Heterorhabditis bacteriophora, H. megidis

and

FAMILY: STEINERNEMATIDAE

Steinernema carpocapsae,
S. feltiae, S. glaseri, S. kraussei,
S. riobrave, and *S. scapterisci*
These nematodes are minute, nonsegmented organisms that are free living in peat and soil compost and proactively seek out their prey. They are a milky-white in color and almost impossible to see when in soil, but when extracted in water they may be seen with basic magnification as individual S-shaped worms (Figure 5.39). The named species are a selection of commercially available nematodes used against a wide range of pests. The list of nematodes and target range of pests is continually increasing.

These species are commercially available in many countries, but a license is required for their use in some.

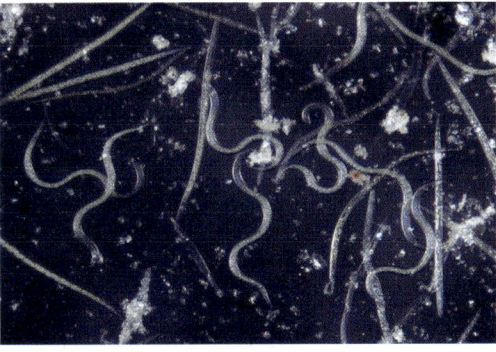

Figure 5.39   Entomopathogenic nematodes (*Heterorhabditis* spp.).

- *Life cycle:* Once a host insect larva is found, the nematodes enter the body through natural openings. (*Heterorhabditis* spp. can also penetrate the insect cuticle.) These nematodes have symbiotic insect-pathogenic bacteria within their gut. *Heterorhabditis* spp. are associated with *Photorhabdus* spp. of bacteria that, in the latter stages of infection, give a reddish color to the host larva (Figure 5.40). *Steinernema* spp. have *Xenorhabdus* spp. bacteria and infected hosts do not change color (Figure 5.41). Soon after host penetration the bacteria are released and, under suitable temperature conditions, multiply, spreading rapidly throughout the insect, causing septicemia and death, usually within 48 hours. The nematode

Figure 5.40   Vine weevil larvae (*Otiorhynchus sulcatus*) infected with entomopathogenic nematodes (*Heterhabditis megidis*); note red/brown coloration.

Figure 5.41   Vine weevil larvae (*Otiorhynchus sulcatus*) infected with entomopathogenic nematodes (*Steinernema kraussei*).

Figure 5.42   Sciarid fly pupa (*Lycoriella auripila*) infected with entomopathogenic nematodes (*Steinernema feltiae*).

feeds on the bacteria and digested host tissue, reproducing to form thousands of larval nematodes that are released from the cadaver some 10–14 days later.

- *Crop/pest associations:* Entomopathogenic nematodes such as *Heterorhabditis* spp. are mainly used for the control of soil-dwelling pests, including various weevil and chafer larvae. *Steinernema* spp. such as *S. carpocapsae* are used against a range of Lepidoptera, such as codling moth (*Cydia* spp.), and specific Coleoptera, such as the flat-headed rootborer (*Capnodis tenebrionis*) that attacks several fruit trees. *S. feltiae* has a wide range of activity including attacking leaf miner larvae within leaves, scale insects on plants, sciarid fly larvae (Figure 5.42) in soil and in mushroom culture, and thrips on plants and in soil. *S. glaseri* is commercially used in several countries against Japanese chafer grubs (*Popillia japonica*). *S. riobrave* was isolated from desert regions in the United States and can withstand high-temperature and low-moisture environments, where it is used against various citrus root weevils. *S. scapterisci* is increasingly used against mole crickets (*Scapteriscus* spp.) on golf courses in the United States.

- *Influence of growing practices:* Most entomopathogenic nematodes are inactive at temperatures below 12°C (54°F). This limits their use in temperate regions mainly to protected crops where the compost can be maintained at a suitable temperature. They can be used successfully outdoors from late spring to early autumn for control of pests in container-grown plants. However, *Steinernema kraussei* is active down to 5°C (41°F) and is used against black vine weevil (*Otiorhynchus sulcatus*) from early spring to late autumn. These nematodes are highly susceptible to desiccation, particularly when used in containers kept in glasshouses or conservatories. Most pesticides are safe to use with nematodes, except for some soil-applied drenches and those with nematicidal activity. Desiccation is a serious problem, particularly with foliar applications, and care to avoid rapid drying of the spray is required for successful pest control. Similarly, soil applications may fail due to lack of moisture, which inhibits nematode mobility. Conversely, active infective stage nematodes can easily drown if left in nonaerated water for more than a few hours, and the same can occur in water-logged compost or soil. Entomopathogenic nematodes are usually applied as a soil drench and can be used as both preventative and curative treatments.

Phasmarhabditis hermaphrodita

Phasmarhabditis hermaphrodita is a parasitoid nematode first isolated from field slugs in the southern UK (Figure 5.43). Slug nematodes are microscopic at 1097 µm × 7 µm (1000 µm = 1 mm). Their natural habitat is soil where they can survive for long periods in the juvenile stage but only in small numbers. The juvenile stage has no mouth or anus, so it does not feed but survives using fat stored in the bodies. It is the bacteria

Figure 5.43   Late stage of infection by *Phasmarhabditis hermaphrodita* **on gray field slug (***Deroceras reticulatum***).**

Figure 5.45   Slug with swollen mantle infected with entomopathogenic nematodes (*Phasmarhabditis hermaphrodita***).**

Figure 5.44   Gray field slug (*D. reticulatum***) infected with** *Phasmarhabditis hermaphrodita* **showing swollen mantle.**

Figure 5.46   Gray water snail (*Oxyloma pfeifferi***) infected with** *Phasmarhabditis hermaphrodita***.**

(*Moralla osloensis*) associated with the nematodes that are the cause of the host organism's death.

This species is commercially available in many countries, but a license for its use is required in some.

- *Life cycle:* Reproduction can only occur on suitable invertebrate host organism tissue. *P. hermaphrodita* move short distances in the soil or compost to seek out new hosts as long as the medium is moist and not too compacted. When a slug is encountered, the nematodes enter the body through a pore at the rear of the mantle called the pneumostome. They develop from infective juveniles to adults. This triggers the release of the bacteria,

which begin to multiply and spread to the rest of the slug's body. The presence of the bacteria and increasing numbers of nematodes within the affected slug can be seen as a swelling of the mantle that results in its death a few days after infection (Figures 5.44 and 5.45).

- *Crop/pest associations:* P. *hermaphrodita* is most pathogenic against the gray field slug *Deroceras reticulatum* and *Arion* species but can also kill other slugs and some snails (Figures 5.46 and 5.47). Slug feeding is inhibited a few days after infection and subsequent plant damage is rapidly reduced. Dead slugs tend to remain hidden belowground, where they disintegrate after releasing the next generation of infective juvenile

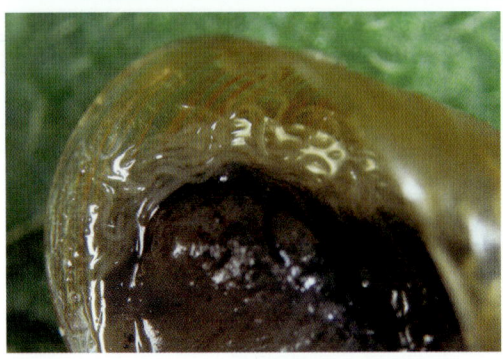

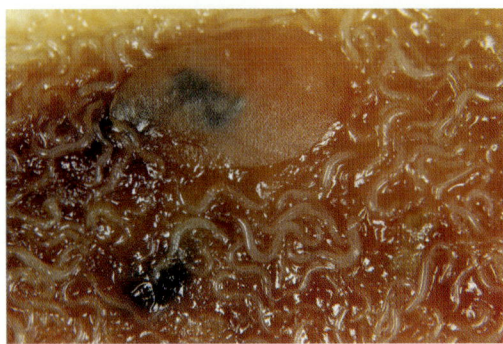

Figure 5.47   Close-up of gray water snail (*P. pfeifferi*) infected with *Phasmarhabditis hermaphrodita* showing free-living stage of nematodes.

Figure 5.48   Free-living stage of *Phasmarhabditis hermaphrodita* swimming in decaying slug ooze.

nematodes (Figure 5.48). This recycling of nematodes does not significantly increase or extend pest control for long periods. Commercial use of Nemaslug (the patented formulation of *P. hermaphrodita*) has increased dramatically and they are now routinely used on several broad-acre crops such as brassicas and potatoes.

- *Influence of growing practices:* Nematodes are most effective in moist soil when the temperature is between 12°C and 20°C (54°F and 68°F). Applications can be made at any time of year as long as the preceding conditions can be achieved (freezing will kill them). Some species of beneficial nematodes, such as *S. kraussei* and *P. hermaphrodita*, have been shown to be active at temperatures down to 5°C (41°F). Natural populations of parasitoid nematodes are usually present in relatively low numbers and do not kill many organisms. However, an application of these nematodes results in a temporary increase, which can provide good control. As the parasitoids do not persist in large numbers for long periods of time, they are unlikely to cause a significant disruption to the ecosystem. Although found in the UK and sold in parts of Europe, this nematode or similar species have not, as yet, been isolated in North America.

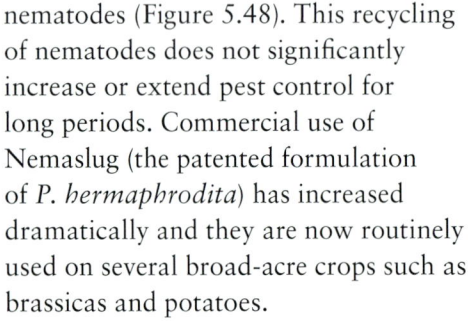

Biological control in perspective

Dr. Mike Copland

WyeBugs, Wye Campus

Imperial College, London

INTRODUCTION

Biological control can be defined as the control of pests or pathogens by other organisms. The introduction of a biocontrol agent is aimed at reducing the population of the pest or pathogen to below an economic damage threshold for that particular crop. But biological interactions occur naturally, and it is through the effects of these that plant and animal populations are regulated and generally kept in balance (Figure 6.1). Our agricultural practices that produce vast monocrop areas distort this balance in favor of the crop plant, which may increase the attack by host-specific herbivores and disease (Figure 6.2). The introduction of bio-control agents is used to reduce the serious-ness of these attacks and maintain a plant population at a higher density than would occur in the wild. Biocontrol is an exercise in population management and with it goes the responsibility of understanding the possible effects on nontarget populations. In this chapter some of these interactions are

considered and the possible mechanisms of biocontrol discussed.

CLASSICAL BIOLOGICAL CONTROL— FOREIGN PLANTS

Our increasing ability to cross the world in less than a day and source our food from faraway places means that some animals and plants that have evolved separately on each continent in different ecological communities for some 60 million years can be introduced into new habitats. One possible consequence of this is that the introductions can cause a major problem because the natural agents that regulate them are not present. There are a number of well documented examples of problems resulting from the introduction of plants that have subsequently been con-trolled by the use of imported biocontrol agents.

An ornamental prickly pear cactus (*Opuntia stricta*), originally from South America and

Figure 6.1   Diverse planting of vegetables in an English allotment.

Figure 6.3   Flowering St. John's wort (*Hypericum perforatum*)—an invasive plant in the US.

Figure 6.2   Fields of flowering oilseed rape (*Brassica napus*) grown as a monocrop in England.

Figure 6.4   Flowering Japanese knotweed (*Fallopia japonica*)—an invasive plant.

taken by settlers to Australia, escaped onto profitable grazing land and, between 1900 and the late 1920s became vast forests of impenetrable prickly jungle. This was only controlled when reunited with a small pad boring moth (*Cactoblastis cactorum*) from its home country. The moth and cactus are still in Australia, but in general, do not pose a problem. Periodic outbreaks can occur— for instance, when flooding carries cacti to areas where the moth has died out or soil disturbance allows germination of long buried cactus seed.

A similar outbreak of the shade-loving European St. John's wort (*Hypericum perforatum*) (Figure 6.3) taken to North America became a dominant, toxic plant in sunny, open grazing land. When reunited with

its European control, a small leaf eating beetle (*Chrysolina quadrigemina*), the plant retreated to a minor status in damp, shady areas.

In the UK, Japanese knotweed (*Fallopia japonica*) was brought in during the mid-nineteenth century as a garden plant (Figure 6.4). It reproduces vegetatively by underground roots that survive chopping into small pieces. Over the last 40 years it has become a major weed, now covering land in the southwest and other parts of the UK, and is costly to control. The year 2010 saw the first trial releases of a Japanese psyllid insect, *Aphalara itadori*, that might bring about its control (see p.107). This work by CABI (Center for Agricultural Bioscience International) has taken over 10 years of risk assessment study to show that the release of this new insect will not harm any other related native plants,

Figure 6.5   Flowering Himalayan balsam (*Impatiens glandulifera*)—an invasive plant.

Figure 6.6   Adult cardinal ladybird or vedalia beetle (*Rodolia cardinalis*) feeding on cottony cushion scale insect (*Icerya purchasi*).

although whether it can control Japanese knotweed remains to be seen.

Other alien weeds occur in the UK, particularly *Rhododendron ponticum* and Himalayan balsam (*Impatiens glandulifera*) (Figure 6.5); so far, these have not been controlled biologically.

CLASSICAL BIOLOGICAL CONTROL—FOREIGN PESTS

In the preceding cases, herbivorous insects were imported to control pernicious alien weeds. However, more commonly, foreign insects attack introduced crops. For instance, citrus species originated in Southeast Asia with closely related species in Australia and have been cultured and spread across the world by man. The Australian cottony cushion scale (*Icerya purchasi*) is a citrus pest accidentally introduced to California around 1868. By 1887 it had nearly succeeded in wiping out the citrus industry. In late 1888 an introduction of a ladybird beetle, *Rodolia cardinalis* (Figure 6.6), from Australia completely controlled the scale insect population back to a nondamaging level (see p.126). Periodic resurgence can occur from the overuse of pesticides in a particular area when the ladybirds' numbers are depleted.

Cassava (tapioca), a South American plant producing underground storage roots (Figure 6.7) and capable of great drought

Figure 6.7   Fleshy storage roots of cassava (*Manihot esculenta*).

resistance, was taken to Africa a couple of hundred years ago and has become the staple diet of huge numbers of people. In the 1970s a cassava mealybug, *Phenacoccus manihoti*, was accidentally brought from South America in new plant-breeding stock and by 1980 was threatening destruction of African cassava and potential starvation of those dependent on this plant. After much searching, a suitable parasitoid (*Apoanagyrus lopezi*) from South America was bred in vast numbers, dropped by air over the cassava growing regions in Africa, and, in most cases, controlled the mealybug.

In Europe, recent insect invaders include the North American Western flower thrips (*Frankliniella occidentalis*) that spread worldwide during the late 1970s and has become widespread in most protected crops.

In the UK, it appears to have few natural enemies and, while some predatory mites and insect bugs are sold to give control, specific parasitoids have not yet been licensed for release in the UK or many other countries.

The horse chestnut, *Aesculus hippocastanum*, is a tree from the mountains of southeastern Europe that has been introduced into many countries. A leaf miner (*Cameraria ohridella*) first observed in Greece in the 1970s is now browning the foliage of horse chestnuts throughout the South of England (Figures 6.8 and 6.9). There are worries that the introduction of a parasitoid control agent may attack other native Lepidopteran leaf miners and compete with other leaf miner parasitoids. Any such release would need to ensure the specificity of the introduced parasitoid for the chestnut leaf miner so that biodiversity is not reduced as a result of its introduction.

Many European gardeners complain that their exotic Asian lilies are now attacked by the Asian lily beetle (*Lilioceris lilii*) (Figure 6.10)—also known as red or scarlet lily beetle (Figure 6.11). This beetle first established in the UK in the 1940s in Berkshire. It remained isolated there for 50 years, but since the 1990s has spread rapidly throughout southern England. Some of its parasitoids now occur in

Figure 6.8   Horse chestnut (*Aesculus hippocastanum*) damaged by leaf miner (*Cameraria ohridella*).

Figure 6.9   Late instar larva of horse chestnut leaf miner (*Cameraria ohridella*) excavated from within leaf.

Figure 6.10   Severe feeding damage on Asian lily caused by larvae of *Lilioceris lilii*. The larvae cover themselves with their own frass, making them distasteful to most predators.

Figure 6.11   Adult Asian, red, or scarlet lily beetle, *Lilioceris lilii*.

the UK but, so far, have had little impact on the pest.

The North American Colorado beetle (*Leptinotarsa decemlineata*), an important pest of the introduced South American potato plant in many countries; the American serpentine leaf miner(*Lyriomyza trifolii*); the Asian whitefly (*Bemisia tabaci*); and, most recently, the South American tomato leaf miner (*Tuta absoluta*) are all visitors to the UK but have not yet become established parts of the fauna. While all of these could be controlled by the introduction of suitable predators or parasitoids, quite a lot of work is required to show that importation of a particular exotic beneficial to control these pests will not harm our native ecosystem.

FOREIGN PREDATORS THAT MIGHT DO HARM

Already we have good examples of imported predators that are doing harm. The Far Eastern harlequin ladybird (*Harmonia axyridis*) was bred and sold for aphid control by a French producer to amateur growers and invaded the UK in large numbers in 2004. In just a few years, it has appeared all over southern England and appears to be displacing several of our native ladybird species.

The south European predatory bug *Macrolophus pygmaeus*, released into the UK under license into glasshouses for whitefly control, has escaped and established in southern England. It has also been reported to cause plant damage in tomato crops toward the end of the growing season. More recently, another mired bug, *Nesidiocoris tenuis*, has been released in Spain for control of the leaf miner, *Tuta absoluta*, and whiteflies, but it is now reported to be causing severe plant damage by its feeding and oviposition punctures.

The North American predatory mite *Amblyseius californicus* was licensed for use in glasshouses in the UK on the basis that it would not survive outside. However, it is now successfully established in the south, where it may be competing with other native predatory mite species.

These examples illustrate how control of plant or insect populations can go awry as a result of the use of biocontrol agents.

ANIMAL GROUPS

At the time of the first land plants, the marine animal groups had proliferated into some 30 different phyla with bodies built with various layers of cells, single or repetitive segmented bodies, internal or external skeletons, and assorted blood systems. Each phyla had developed into carnivorous, herbivorous, filter feeding, free swimming and bottom living species. Only a handful of these phyla subsequently developed on the land, but each has representative plant-feeding, carnivorous, and saprophagous species that live on dead and decaying material.

Animals that feed on land plants comprise the invertebrate nematode worms, slugs and snails, and the arthropod group including mites, insects, some woodlice, and symphylids. Many of these are considered to be pests, although when they are used as biocontrol agents they are called beneficials. Grazing vertebrate feeders include rabbits; single-hoofed grazers like horses; cloven-hoofed grazers like sheep, cow, and deer; and some birds. Many vertebrate families (birds and mammals) are involved in seed dispersal as a result of eating fruits.

Nematodes

Nematodes (eelworms) include an economically important group of plant and animal parasites. They are microscopic in size, but responsible for a host of serious animal and plant diseases that form the basis of many of the restrictions for moving plants, animals, and soils around the world today. Nematodes are thin, non-segmented, worm-like animals mostly less than 0.5 mm in length; they are unusual because they have a gut without any kind of digestive enzymes. Food is moved through the gut by a muscular pump just behind the head and nematodes, like insects, have an exoskeleton that is molted at intervals as they grow. They feed on readily absorbed simple sugars and amino and fatty acids and therefore live among colonies of bacteria or have a means of specifically puncturing cells and feeding on the cytoplasm within.

Plant-feeding nematodes: The plant-feeding nematodes fall into a number of categories. Some simply live in the soil and puncture and feed on plant roots, and some spread virus diseases between plants. Of greater importance are the endoparasitic species that live within the plant leaf (Figures 6.12 and 6.13), stem, bulb, and root, with the latter sometimes having specific cyst and gall species for which the plant provides appropriate nutrients. Plant-feeding nematodes appear to be an ancient group

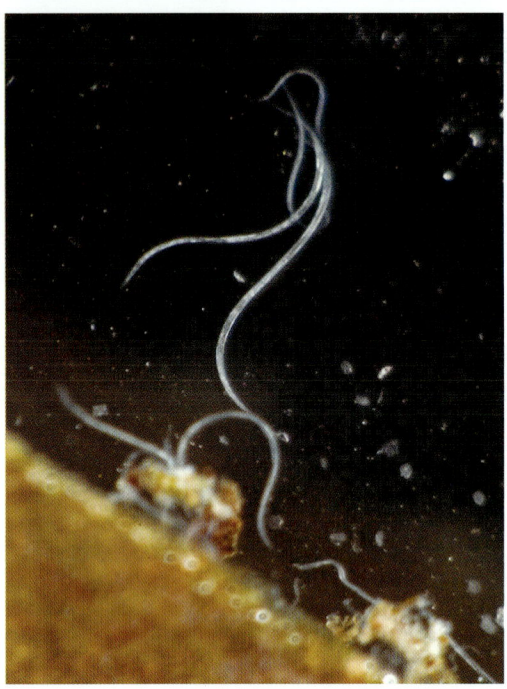

Figure 6.13   Nematodes oozing from leaf edge when above leaf was placed in saucer of water with a drop of detergent prove the presence of active leaf nematodes.

found throughout the world that attacks many families of mainly flowering species. Some plants produce nematode antagonist chemicals (e.g., marigolds [*Tagetes*]) and most nematodes species have a narrow and specific host plant range. Nematodes have no obvious means of widespread dispersal, yet they have been carried around the world with their host plants and are a major problem. Recognizing and destroying infected plants is vital for nematode control in both the garden and the field. Resistant egg or larval stages can be carried by the wind, in water run-off, and in the soil carried on animal feet, and, in some cases, are vectored by insect herbivores. Plant-attacking nematodes play an important role in the control of plant populations; potatoes and sugar beet are examples of UK crops that can be devastated by cyst nematode attack (Figures 6.14 and 6.15).

Figure 6.12   Underside of Japanese anemone infested with leaf nematode (*Aphelenchoides* spp.). Note angular pattern of damage and lack of spores that may have indicated downy mildew.

Figure 6.14   Potato crop devastated by potato cyst nematode (*Globodera rostochiensis*).

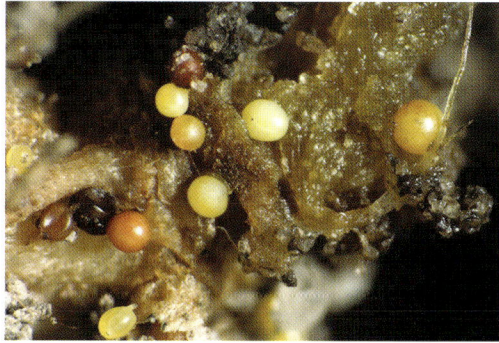

Figure 6.15   Cysts of *Globodera rostochiensis*; this stage can remain dormant in infected soil for up to 20 years.

Animal-feeding nematodes: Many nematodes are of medical and veterinary importance and it is common for humans to have nematodes encysted in their tissues. Some nematodes are used in biological control (see p. 201). For instance, species in the genus *Heterorhabditis* are used commercially for control of soil pests like vine weevil, *Steinernema* for control of soil and foliar pests, and *Phasmarhabditis* for slug control. A suspension of nematodes can be placed in water and sprayed onto plants or applied to the soil. The nematodes are able to enter the host through natural openings (mouth, anus) and, once inside, release specialist bacteria that take over, kill and liquefy the host, rather like a spider liquefies the body of its prey, and then suck in the digested fluid. The nematodes then multiply

inside and when they are mature leave the body to seek new hosts carrying their lethal bacteria.

Mites

The mite group comprises many cell feeders that are important in both plant population control and in plant protection by attacking plant herbivores. Most gardeners and horticulturalists will have encountered plant-feeding mites. The glasshouse spider mite in the family Tetranychidae is typical. These small creatures have mouthparts that suck the cell contents and increase rapidly, often killing the plant. A related family of Tarsonemid mites often lives concealed in developing plant buds, causing distortion of subsequent leaves and flowers (Figure 6.16). Eriophyid mites are a family of minute mites that live within the plant tissue (Figure 6.17) and, in some species, stimulate the plant to produce blister-like gall structures such as the red-tinged galls of *Eriophyes erineus* seen on walnut leaves (Figure 6.18), as well as the familiar witches'

Figure 6.16   Black currant big-bud or gall mite (*Cecidophyopsis ribis*) damage to young leaves.

Figure 6.17   Beech gall mite (*Eriophyes nervisequus fagineus*) causes minute galls on leaf hairs.

Figure 6.18   Walnut leaf gall mite (*Eriophyes erineus*) damage.

brooms on trees and enlarged buds on black currants (see p. 38).

Many mite families, such as blood-sucking ticks and chicken red mites, attack animal groups. Skin mites cause scabies and various mange conditions. Some of these are important vectors of animal virus diseases.

In biological control, a range of mite species is now used as important predators for familiar plant pests. The family Phytoseiidae, for instance, has members that control spider mites, thrips, and whitefly. The Laelapidae family contains an important genus (*Hypoaspis*) that is mass produced for the control of small soil pests such as sciarid flies. Not all predatory mites are seen as beneficial. The family Varroidae includes the Far

Eastern varroa mite, which attacks honeybees and has spread worldwide over the last 40 years (Figure 6.19). There are thousands more species of predatory mites, and many are currently being developed commercially for biological control purposes.

Many texts refer to mites being dispersed by winds. Mites have no wings of their own, but many can certainly hitch lifts with flying insects (phoresy) that pollinate (Figure 6.20), feed on, or protect particular plant species by homing in on specific plant volatiles and even with other mites (Figure 6.21). Ladybirds in the genus *Chilocorus* carry mites on the underside of their thickened forewings to the scale insect prey, which provides food for both the beetle larvae and the mites. The mite

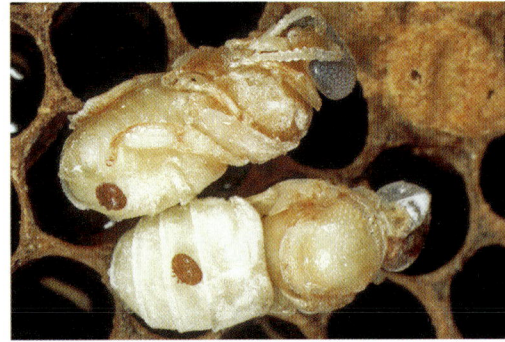

Figure 6.19   *Varroa destructor* is an ectoparasitic mite of the Western honeybee (*Apis mellifera*), seen here on the pupae.

Figure 6.20   Mites on bumblebee (*Bombus terrestris*) showing phoretic behavior.

Figure 6.21 **Phoretic mites on predatory mite** (*Macrocheles robustulus*).

and beetle compete for the same food, but act together as effective biocontrol organisms.

Our modern flowering plants have evolved with a much more important group of arthropods—namely, the insects. The relationship began in the Carboniferous period some 300 million years ago. Today, insects are responsible for regulating plant populations as well as giving protection from herbivores, pollinating flowers, and vectoring various pathogens.

MECHANISMS INVOLVED

Biological control is not just about reuniting species with their natural population regulators, but also about understanding the mechanisms involved in determining population control, plant protection, and pollination. Every plant species has evolved as a member of a natural plant community and, under similar climatic conditions, will develop a stable population at a particular density per unit area. At low densities, plants may avoid herbivores and disease, but pollination may be more restricted. At high densities, pollination is generally effective, but herbivory and disease may also be high.

All plants give off volatile signals and, when the plant population in a specific area is high, herbivore species are more able to locate them and feed, thereby reducing their numbers. The density per unit area where the ecological balance is stabilized will vary with the plant species and the situation. So, different species of plants may have a density measured at so many per square meter, while others may stabilize at so many to the square kilometer.

However, this applied to natural habitats before the intervention of man and includes complex forest communities to more open grassland, sand dunes, and salt marsh habitats. During the geological Cenozoic period, when our flowering plants were developing along with pollinating insects, mammals, and birds, the continental land masses were moving toward their current dispersed position with a cooling climate. Some of our temperate trees developed a deciduous habit, with wind-pollinated flowers emerging before the new spring leaves—thus avoiding the insect pollination density requirement. Similarly, the grasses (Poaceae) originating in the Cretaceous with dinosaurs now developed as an important wind-pollinated family with a specialist basal growth habit that supported the feeding and development of grazing horses and cloven-hoofed mammals (Figure 6.22). Without grazers, grasses are rapidly outcompeted by broad-leaved plants. These large grazers lack the specialist discriminatory senses of

Figure 6.22 **Scottish blackface ewe grazing on lush meadow grass.**

insects and maintain the grass-dominated habitat. Wild grazing animals tend to be gregarious and somewhat nomadic, so there is still an opportunity for both the grasses and the broad-leaved plants to flower and for insects to play a density-controlling role. The periodic returning grazers prevent the grassland habitats from reaching a natural forest climax.

Volatile signals are detected by specialist sense organs on the insect's antennae, each of which is stimulated by particular chemical molecules. A particular plant species will have a "signature" natural scent comprising a mixture of several different volatile molecules in different ratio or proportions. A related plant species may give off the same molecules but in a different ratio. A specialist insect herbivore species can detect the ratio; it reacts to one plant by flying toward it, while having no interest in the other plant.

This is a sophisticated communication system between plants and all the animals with which they have coevolved—whether they are leaf-eating, pollinating, or fruit-eating insects or mammals like bats. It is a language similar to what we use, in which the chemical molecule is the equivalent of the sound of a letter of the alphabet, and the numbers of different molecules spell words of different lengths. The ratio of each molecule produces an emphasis on certain components to which the insect responds, just like human spoken language. These signature scents vary throughout the season with the growth of the plant and the plant organs that produce them. Flowering, fruit production, and disease all alter the profile and therefore the attraction or repellency of the plant. Unlike human language, there is an unlimited number of letters to the alphabet so that a complex plant community made up of hundreds of species can potentially all communicate simultaneously with their insect partners without confusion. In practice, however, the pro-

duction of volatiles from different plants may vary at different times throughout the season, at different times during the day or night, and also in response to temperature, light, humidity, and water stress in a way that coincides with the availability of the insect helpers.

When the herbivores find the plants, the damage done by feeding produces new plant volatile signals that combine with the herbivore's own volatiles (e.g., saliva, frass [faeces]) and, in some cases, the herbivore sex pheromones). These may attract specific beneficial predators or parasitoids that attack the herbivore. This ability of herbivore insects to react to the volatiles of a particular plant—or a predator and parasitoid to recognize the volatiles from a damaged plant—are all encoded in their genes. However, to fine-tune this process, we also know that new adult insects on emergence are influenced by the smell of their host and plant species on which they developed. There are some possibilities of exploiting this smell memory during commercial rearing to make the parasitoids more likely to work in particular crops. There is also the possibility of producing a new generation of harmless but effective plant protection chemical lures that give off these calling signals to encourage particular beneficials to visit the crops before damaging pest populations have developed. Early work shows a higher level of parasitism where such lures are used.

Similarly to plants, a single caterpillar or aphid will have a poor volatile profile, but as numbers and size increase, the signal strength gets greater. These volatiles can call in beneficials from a greater distance, so excessive herbivore action is brought under control and the plant population somewhat protected and stabilized.

Both predators and parasitoids also have their own controllers in the form of further species of parasitoids and hyperparasitoids, respectively. Of course, in a natural habitat there will be hundreds of species of plants: two to three times as many specialist herbivores and perhaps as many again of predatory beneficial insects. So much chemical communication going on in a habitat needs to be regulated. In practice, these signals tend to be arranged seasonally, as herbivore activity coincides with flushes of plant growth, fruiting, or flowering. As such, many of the insect herbivores, their beneficial controllers, and, in turn, their controllers may have only a single generation a year and a narrow period when they are seen. A beneficial insect that has a fixed, obligatory one generation a year can be perfectly satisfactory in a natural environment, but may be of little use as a biological control agent where many generations a year might be required to control an agricultural pest.

A good example of this complexity is the cabbage plant, now grown worldwide and attacked by a variety of greenfly, whitefly, and assorted moth and butterfly species. One of these is the large cabbage white butterfly, *Pieris brassicae* (Figures 6.23 and 6.24). Across the world, many species of parasitoid have been recorded attacking this single pest species. Here in the UK, the eggs are attacked by a tiny parasitic wasp (*Trichogramma brassicae*) (see p. 179), which is attracted to the sex pheromones of recently mated females and hitches a lift with a female to the next cabbage, where both the butterfly and the parasitoid lay their eggs. The parasite can lay up to three of her eggs into each butterfly egg. Uninfected eggs successfully hatch into small caterpillars and give off volatiles that call in a small parasitic wasp (*Cotesia glomerata;* see p. 167), which takes only a couple of seconds to inject up to 100 eggs into the caterpillar body. These eggs hatch into tiny grubs that then wait until the caterpillar is nearly fully grown. At this point they devour the caterpillar from inside and break out to produce bright yellow cocoons around the body of their victim (Figure 6.25).

Hyperparasites (particularly *Tetrastichus sinope*) seek out infected caterpillars and lay their eggs through the caterpillar skin into the developing parasitoid grubs.

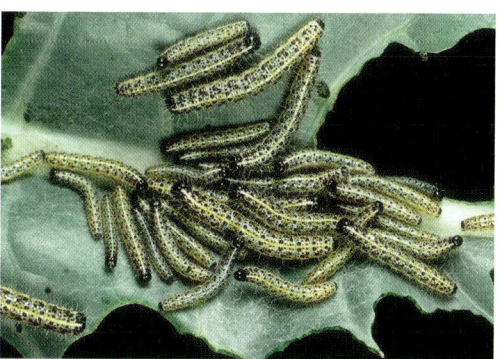

Figure 6.24   Third and fourth instar large cabbage white butterfly (*Pieris brassicae*).

Figure 6.23   Neonate larva of large cabbage white butterfly (*Pieris brassicae*) with empty egg cases.

Figure 6.25   Larvae of parasitoid wasp (*Cotesia glomerata*) emerging from dead cabbage white butterfly (*Pieris brassicae*).

MECHANISMS INVOLVED

Several other species of parasitoid attack the yellow cocoons that form around the dead caterpillar body. Caterpillars that were not attacked when young are also vulnerable to attack by a large parasitoid, *Pimpla turionella*. Once safely turned into the pupal stage, the butterflies are now attacked by a small, metallic-green parasitoid, *Pteromalus puparum,* that lays over a hundred eggs at a time. While the cabbage white butterfly does an amazing job of destroying the gardener's cabbages, most of the parasitoids come in too late to save the damage to the plant, with only the egg parasitoid being successful. Overall, very few caterpillars succeed in completing development to the adult butterfly, with around 95% parasitism being common.

However, if the cabbage is attacked by cabbage moth *Mamestra brassicae* or the diamondback moth *Plutella xylostella,* none of the previously discussed parasitoids are interested, so another group is called into play. Similarly, if the cabbage is attacked by mealy cabbage aphid, then the signals bring in a different parasitoid that specializes in that aphid species, while cabbage whitefly attack again produces its own group of specific parasitoids.

Pollination and Seed Dispersal

The process of evolving a population control mechanism that determines the density of the plant also has an important impact on the mechanisms of pollination and seed dispersal. Various animal groups are involved in this process.

Flowers use scent, color, and nectar to attract their pollinator insects. They also use complex chemical molecules to repel most and attract only specialist pollinating species. In this way all plants ensure that only specialists will acquire their pollen, which will be taken only to other plants of the same species. Plants that normally grow at a low density often have brighter colors, larger flowers,

Figure 6.26   Swallowtail butterfly (*Papilio machaon*) taking nectar from flowers.

and stronger scents on tall inflorescences and require pollinators that fly long distances, such as the more recently evolved strong flying groups like butterflies (Figure 6.26), moths (from the Paleocene period 65 million years ago), and bees (from the Oligocene period 35 million years ago); they have long nectar feeding proboscises.

The butterflies and moths have a leaf-eating caterpillar stage that helps to regulate plant populations, but an adult can only feed on nectar. The feeding process enables the pollination of plants of a different species that may live either in the same plant community or in an adjacent community.

Bees, on the other hand, visit plants to feed on nectar but also collect the pollen grains to use for rearing their young (Figure 6.27). In collecting the nectar, they are at the same time carrying out a pollination process. In this way the same plant is used to feed both the adult and the bee's young larvae.

This group of insects requires specific diet mixtures of complex sugars, amino acids, and perhaps lipids, and the flowers have to become specialist tube shaped structures that prevent entry by most insect species but reward their pollinators with the correct food to develop their eggs and, in some cases, specific chemicals needed in their metabolism. This is a lock and key mechanism linking plant and pollinator together.

Figure 6.27   Adult honeybee (*Apis mellifera*) with pollen on legs collected from flowers.

Figure 6.28   Soft scale insects such as *Coccus hesperidum* produce copious quantities of sticky honeydew.

Sugars That Fuel Animal Activity

The most important aspect of making biological control work is to provide the beneficial insects with a competitive advantage, such as shelter from rain, sun, and wind and provide sustenance to develop their eggs and energy for flight. Some of the sustenance can come from flower nectars, and these beneficial insects are active pollinators as well as pest controllers. However, in many cases the energy source comes from other insects. For instance, the rather despised Hemiptera group includes the sap-sucking aphids, whiteflies, mealybugs, and scale insects (Figure 6.28) (Sternorrhyncha or Homoptera) that probably originated well before the evolution of insect-pollinated flowering plants from the Jurassic period (130 million years ago).

This sap-sucking group has specialized mouthparts that pierce the plant surface and delve between the cells to the deeper sap-carrying phloem tubes. This part of the plant's vascular system is responsible for distributing the sugars made in the plant leaves by the process of photosynthesis. The phloem cells are under pressure and, once pierced by the insect's stylets, the sugary sap pumped by the plant flows to the sucking insect. This sugar passes through the insect gut, where some nutrients are removed.

During this process the sugar is transformed by various enzyme systems into a variety of more complex sugar types. When mixed with plant volatiles, it produces unique honeydew that in turn becomes the food for particular groups of insects. Therefore, on each plant each specialist sap sucker produces its own unique blend of flavored sugars that will in turn feed the specialist insects that benefit the plant. In some cases the presence of a particular aphid or mealybug colony acts as a nutrient sink equivalent to the development of floral and seed structures. The plant appears to respond by directing more nutrients to the affected parts. These microhabitats often support several species of sucking insects in a communal relationship. It can be shown experimentally that the presence of a mealybug may increase the rate of growth of a colony of soft scale insects.

A plant (whether it is flowering or not) with a colony of aphids or mealybugs will now have a supply of external sugars that will feed not only species that keep the aphids under control but also parasitoids and predators of other herbivores such as caterpillars. Of course, not every species in a mixed plant community needs its own sucking insects. But every crop must have support for beneficial insects and, in many cases, this may be a scattering of noncompetitive "weeds" that support the sap-sucking insects. Today, some farmers provide specific "beetle banks" of rows of suitable plants at, say, 100 m intervals from which the predators can forage into the crop. Field margins are recognized and managed as a potential rich source of nectars and nonpest sucking aphids with their supplies of honeydews. In addition nonaggressive weed cover and companion planting help beneficial insects survive within the crop itself.

The Combination of Sucking Insects and Ants

The honeydews produced by the sucking insects are not just a food for beneficial insects, because one of the early groups to use hemipteran honeydew was the social ants, with the first fossils turning up in the early Cretaceous period (100 million years ago). Most ants are carnivorous, but sugars play an important role in

Figure 6.29   Red ant (*Myrmica rubra*) attending colony of plum-thistle aphids (*Brachycaudus cardui*) feeding on globe artichoke.

their lives by providing a compact form of energy. In every natural plant community, ants now play a major role in regulating the densities of sucking insects and plant pests (Figure 6.29). The honeydew produced by sucking insects is a powerful attractant for beneficial insects, including their own natural enemies like ladybirds. To avoid attracting their own controllers, the sucking insects store the honeydew produced during the day inside the rectal region of their gut. At night, when fewer predators are about, they squirt the honeydew as far away as possible from their body so that it falls on the lower leaves or ground beneath.

However, in most natural ecosystems the ants tend the colonies of sucking insects both night and day, and by bringing their mouthparts to the anal region and stimulating the insect with the antenna, the honeydew is ejected into the ant's mouth to be taken back to the nest. The ant in turn destroys enemies of the sucking insect as well as caterpillars and other pests on the plants.

The ants may also carry the sucking insects to other more suitable parts of the plant—rather like a farmer moving his livestock to better pastures. If the aphid numbers get too high, there will be insufficient ants to look after them; the honeydew is now released and its calling signals attract natural enemies that cull the surplus suckers. The net effect is that a plant with a population of sucking insects is a plant that can feed beneficial insects and attract ant attention that, in turn, keeps the sucking insect population controlled and the plant free of other pests.

Development of Resistance to Pesticides and Pollination Drives the Need for Biological Control

There is public concern about possible damage to human health from pesticides. The development of resistant strains of pests has required the continual development of new pesticides and recognition of the possible environmental damage they cause. Today there is no doubt that the real driving force for encouraging a biological approach comes from the need to protect and promote pollinating insects. In protected glasshouse crops such as tomato, raspberry, and strawberry, the huge financial advantage of using bumblebees for pollination (Figures 6.30 and 6.31) has been used alongside the use of biological control for the numerous pest species and thus has avoided the use of insecticides that might harm the bees.

Figure 6.30 Commercially produced bumblebee hive used to pollinate protected raspberry crop.

Figure 6.31 Inside hive, showing bumble bee (*Bombus terrestris*) colony with developing cells. The tapered blue tube is designed to allow bees in but not back out; this is used before a crop spray to "lock" the bees inside.

REGULATION OF BIOLOGICAL CONTROL

In the past, introducing foreign biocontrol agents was considered a sensible practice. Today, however, regulation of biological control is now seen to be of great importance in preserving the natural native species in a country. In the UK, the 1981 Wildlife and Countryside Act, while allowing predatory beneficial insects to be imported for research, made it an offense to release them into the wild without a license. Then, in 1993, the Plant Health Order made it an offense to bring into the country any non-native plant-feeding insects, even for research purposes, without a ministry license.

Curiously, and potentially wrongly, bringing non-native plants into the country is still allowed, with certain exceptions. This trade in exotic plants is probably greater than ever before and brings with it the possible introduction of foreign pests, pathogens, and beneficials. One reason for this is that plants from another continent are less likely to be attacked by native pests and diseases, so they do rather better as crops or in gardens than native species. But this is the very characteristic that may cause some to become serious and uncontrollable alien weeds. Those that are attacked by native pests give off chemical signals that may not be recognized by native beneficial insects.

Recently, the incidence of new foreign diseases attacking our foreign and some UK plants has been seen. *Phytophthora ramorum* and *P. kernoviae*, first seen in the UK in 2002 and 2003, respectively, attack a diverse range of ornamental shrubs and trees such as rhododendron, camellia, magnolia, and horse chestnut. Ash dieback disease caused by the fungal pathogen *Chalara fraxinea* causes leaf loss and crown dieback in affected trees; it was originally found in Poland in 1992 and usually leads to tree death. By late 2012 it was found in established woodland in most counties of the UK. Perhaps one explanation is that these are diseases that have no coevolutionary history with their host plants. Secondly, the plants may lack the protection and stimulation of natural fungal defenses that they would have had in their home country. Thirdly, new virulent diseases result from recombination of genes from several fungal species or strains. This is a worrying development with no obvious solution apart from the destruction of all infected plant material.

A major problem is that some of the plants we grow have tall and colorful scented inflorescences and, with large aromatic and succulent fruits, are the very plants that grow at low densities in natural habitats. As such, they have had to develop strong allurement volatiles to bring in their pollinators, their seed dispersers, their population regulators, and their beneficial protectors. As a result, the dense populations of foreign plants that the gardener or agronomist uses are particularly vulnerable to attack when their natural controllers are accidentally introduced. Good examples are the potato crop, which can be devastated by blight (*Phytophthora infestans*), potato cyst eelworm (*Globodera rostochiensis*), and lilies by red or scarlet lily beetles (*Lilioceris lilii*).

When we grow foreign plants, their natural pollinators are also absent; however, the social honeybee, with its ability to learn to collect pollen and nectar from plants it has never encountered before, enabled these new foreign plants to be pollinated and set fruit. This ability has turned the honeybee into a unique, all-purpose pollinator with the advantage for the beekeeper of producing a benefit of its own in the form of honey. Maintaining fit honeybee colonies is a major objective and concern for the pollination of foreign crop plants that we all rely on. Foul brood disease has

been a problem for beekeepers for many years and, all over the world, beekeepers have been confronted by the varroa mite (*Varroa destructor*) that spread from the Far East in the late 1970s. Today, we worry about bee decline sickness, which may be a new virus disease—vectored by the varroa mite or perhaps low-level contamination by ever more efficient systemic chemical pesticides like some neonicotinoids.

Worldwide, the population of honeybees is now greater than ever before. All bees collect both pollen to raise their young and nectar to fuel their flight, and they have the ability to remember the journey path from the flower source and the nest site. A social honeybee returning from a successful forage trip passes on to other bees the floral scent to look for and, using a unique dance, indicates the direction and distance to fly to find the flowers. The intelligence of honeybees enables them to learn how to deal with foreign plants and they consequently have become the obvious insects to use for pollination of foreign crops. For many humans, honey, which is stored nectar that enables the nest to survive through adverse periods, is the main reason for domestication. I suppose one may view the average beekeeper as a "Fagin" (from *Oliver Twist*) character that uses an army of bees to steal the nectars from neighbors' gardens to stack away honey for financial gain. That may seem rather harsh, but it is certainly true that honeybee hives rapidly deplete local nectar sources and may well be a major player in the declining populations of bumble- and solitary bees, butterflies, moths, and other nectar-dependent insects.

It should be remembered that flower production is an essential step toward seeds. For many tree species—particularly fruit trees—the number of flowers produced in the spring is dependent on bud-set the previous year. Since all the flowers appear

Figure 6.32 Honeybee hives in field of canola or oilseed rape (*Brassica napus*).

simultaneously, suitable weather conditions and the presence of pollinators at flowering time are essential. For pod-bearing plants, such as beans and oilseed rape, adequate pollination is also essential to ensure the pod is full of seed. In both cases, good pollination— generally, over a short time period— is required in order to get a good crop (Figure 6.32).

However, most plants produce their flowers over a prolonged period, which maximizes the chance of good pollination. Plants with a typical spike-like raceme inflorescence of multiple flowers (from foxgloves to complex compressed daisies) have the potential for an indefinite number of flowers to be produced. Each flower, if unfertilized, retains the petals' color and scent production and remains receptive for a little longer period of time. But once fertilized, that flower drops its signaling petals and starts to produce the seed structure and new, younger flowers open. As more seeds are set, the apical tip that produces new flowers is switched off and the plant now develops just the fruit structure and the seeds within. If you are growing a plant for its attractive flowers, the very last thing you want to see is pollination, and it pays to remove set flowers so that new ones are produced. So the gardener will deadhead old roses and regularly pick young runner bean pods to maintain flower and pod production.

Once the fruit is set and ripe, the plant now signals with two more groups of animals: the birds and mammals. This is done by the production of a succulent fruit with colorful pigments, often in shades of red; the production of strong volatile scents and enhanced sugars and other biochemicals that gives the feeder a burst of energy; an attractive taste; and the desire to eat more. These seeds usually have a tough resistant coating and can pass through the animal gut unharmed to be dispersed far away. Indeed some fruit speed the passage of food through the gut; resulting in better seed dispersal. Others have chemicals that repel the wrong species (Solanaceae).

Some members of the Fabaceae (e.g., *Laburnum* and *Phaseolus*) produce seeds that are protected by bitter and toxic alkaloids. Plums and peaches (Rosaceae) produce a very large nut-like structure that

Figure 6.33 Ripe plum (*Prunus domestica*) cut in half to show large seed surrounded by soft edible flesh.

protects the edible seed inside and is collected by rodents to set aside as winter larders (Figure 6.33). Squirrels have developed a formidable and long-lasting memory and scent detection system to hide nuts in the ground as a larder store, while mice may set aside a more conventional single larder. Since many of the larder keepers are eaten by predators, effective seed distribution is ensured.

The fruits of the Rosaceae family appear to have little chemical protection (suggesting that many animals groups are involved in their distribution) and are usually considered safe for human consumption. The fruits have either small protected seed that can be swallowed or extremely hard and large seeds that are spat out after eating the flesh.

Sometimes the fruit volatiles are enhanced by associated rots or with yeasts, and these may attract distributors of the affected fruits. These fruit-rotting microorganisms and yeasts are usually vectored by specialist insects that may also eat some of the fruit material. Fruit pest species like codling moth of apple, whose larvae consume some of the seed and leave an exit hole (Figures 6.34 and 6.35) when the fruit is ripe, may be part of the distribution process for the apple tree, as affected fruits fall and rot first and may then attract foraging animals such as pigs (Figure 6.36).

Sadly, modern agriculture pays no attention to the mechanisms and the specific volatiles that control plant and insect densities. Many of our modern crop plants have been manipulated by plant breeders to increase their nutritional value with higher sugar and protein contents. When this is done, many plants have their repellent, bitter alkaloid and tannin components removed, together with their calling signals, and are then grown as genetically uniform, identical aged monocultures at densities far greater than have ever existed in a natural ecosystem. While we may argue that this has enabled us to feed a starving world, it is only achieved by the use of protective pesticides.

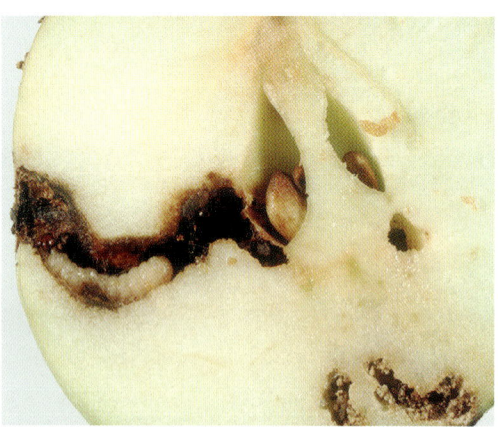

Figure 6.34   Mature apple codling moth larva (*Cydia pomonella*) exposed within feeding gallery inside apple.

Figure 6.35   Exit hole produced by apple codling moth (*Cydia pomonella*).

Figure 6.36   Fallen apples on ground; although called "windfalls," many will have been attacked by codling moth, causing them to dislodge more easily.

In native species, such plants can attract attack by native herbivores, which do extremely well on these enhanced nutritious and resistance-depleted plants; even when beneficials arrive, they are unable to cope with such rapidly expanding pest populations. When one or more of the players have not coevolved with all the others, the signaling system is less than optimum. So a plant in a new environment that is attacked by a pest species not encountered before does not give off the correct volatile spectrum to attract a beneficial that would have attacked it back in its home country. There is no quick fix and it may take a long time for natural selection to evolve a suitable solution.

One of the particularly sad aspects of plant breeding is that all of our cultivated crop plants, our garden flowers, and even our forest trees lack genetic diversity. This means that if an insect or a disease occurs that severely damages a particular cultivar, then sooner or later we must abandon that cultivar and breed a new resistant one. Such methods of plant breeding have been widely used and, for annual plants, these man-made cultivars have been sufficiently inbred over many generations so that they may breed true from seed or are produced as F1 hybrid seed that needs to be repurchased each year. Perennial plants, however, are usually propagated as cuttings and, in the case of tree species, may be grafted on specialist dwarfing and disease-resistant rootstocks. Even our forestry trees propagated from seed are from selected mother plants, perhaps from a very limited parental line. In all cases the genetic diversity within these cultivars is extremely limited, so the appearance of a virulent disease or pest has a rapid and disastrous effect on all plants of that cultivar.

It is here that our synthetic plant lures referred to earlier can certainly help to bring beneficials to protect a crop where natural signals may not be effective. Current work in the UK is looking to use molecular techniques to incorporate genes into cereal cultivars that might repel aphids and attract parasitoids. This push–pull strategy has enormous potential for the reduction of chemical use and the expansion of biological control. Here, a genetic modification (GM) technique has been used to overcome the loss of natural resistance, genetic diversity, and unnatural plant density that characterize our modern crops. It replaces the natural plant properties lost through conventional plant breeding.

The relationship between insects and plants as pollinators and population stabilizers is usually at the macro level. Insects will not generally kill plants; however, they can do far more harm if they carry diseases to the plant or their wounding action produces the opportunity for disease organisms to enter the plant tissue.

MICROORGANISMS

We have known for a long time that microorganisms play a vital role in the life of both plants and animals. Most people will refer to bacteria, fungi, and viruses and associate them with infectious diseases. For many years the agriculturalist would have pointed out the importance of nitrogen-fixing bacteria in the roots of legumes and fungal mycorrhizal root associations in some plants. As a result, our own lives and those of plant growers have been a regime of maintaining cleanliness, disinfection in the house, and sterilization of growing media. On the back of this, a huge chemical industry has focused on pesticides, fungicides, and medicines for the control of plant and animal diseases. The result has been the production of less naturally protected animals and plants and allegations that the chemicals may be harming natural immune systems. Perhaps a better explanation is that immune systems are only developed by challenging them with diseases, coupled with survival of the fittest.

Plants appear to produce a larger mass of food material more economically than can be produced from animals. One of the problems with all plants is that they comprise large cells whose mass is made largely of strong cellulose walls, which most animals cannot digest, and a large, water-filled space that gives the cell its turgidity. Much of this material is of low nutrient value as animal feed and often lacks some essential animal vitamins and amino acids.

One of the features of microorganisms is that their physiology has more in common with animals and their nutrient value is far higher than that of a plant. Consequently, when a plant is invaded by a colony of microorganisms, its nutrient status goes up and this coincides with an increase in a defense system that minimizes the chances of its being eaten. A mildew-infected pasture grass tends to be avoided by grazing animals and insects, whether caterpillars or aphids. Whether these chemical defenses are part of the disease attacking the plant, the plant responding with an attack on the disease, or a response by one or the other to prevent a third party joining in is of great interest. Grazing animals and humans can be poisoned by eating diseased plants—in particular, the fetus in a pregnant woman. For instance, ergot diseases in cereals (Figure 6.37) and potato tubers infected with blight can have very serious effects if eaten. Some of the compounds produced in affected plants have effects on the central nervous system, making the eater more vulnerable to predation.

In general, infected plants become more repellent to insects and other microorganisms, although this may change when the microorganism is ready for dispersal. Many disease life cycles will have some specialist relationships with a vector where the disease produces calling signals for specific animals to develop in the rotting plant material or perhaps land among sporulating structures and then move on, carrying with them infectious material.

Healthy plants' natural defenses can be far more toxic than many man-made

Figure 6.37   Ripe ear of wheat showing ergot disease (*Claviceps purpurea*).

Figure 6.38   Green mold (*Penicillium digitatum*) on ripe orange fruit—one of the storage rots.

pesticides, so it is not always the case that eating pesticide-free food is the best. Part of a microorganism's defense ensures that the plant is also more resistant to other microorganisms. This defense is particularly well seen between bacteria and fungi; the latter group is responsible for some antibiotics (Figure 6.38) that we now use in medicine for control of bacterial disease. Similarly, fungal defense chemicals are seen as the basis of a potential new generation of biopesticides.

More recently, our desire to understand the basis of our genetic code as well as the control of gene expression has been applied to the vast number of microorganisms. The result has been a complete revolution in our understanding of their classification, their inter-relationships, and their possible role in the lives of higher plants and animals.

Bacteria

Bacteria are considered our oldest life forms and some of the cell organelles of higher organisms, such as mitochondria and chloroplasts, are derived from bacteria. These organelles drive the cell processes and are passed on in the maternal line to the offspring. The related *Rickettsia* and *Wolbachia* that cause diseases are considered close relatives. A number of formerly unplaced microorganisms that lack cell walls, like *Phytoplasmas*, *Mycoplasmas*, and *Spiroplasmas*, are now placed in the bacterial class Mollicutes; many infect plants and are involved with insect vectors.

In nearly all animals and including some insects, bacteria in the gut play an important role in food digestion and are responsible for synthesizing materials that are deficient in the diet. In all the sap-sucking insects in the Sternorrhyncha, an endosymbiont bacterium, *Bruchnera* sp., occurs in specialist cells in the body providing specific amino acids to supplement their sugary diet. Feeding insects with antibiotics often kills

them and indicates the importance of the bacterial role.

Wolbachia is found in many animal species, particularly in arthropods. Here, *Wolbachia* is responsible for distorting sex ratios—for instance, by killing males during development (common in ladybirds) or enabling unmated females to produce viable female offspring parthenogenetically (thelytoky). Some of our common pests, like thrips, scale insects, some stick insects, vine weevil, and many parasitic Hymenoptera, are only known as females. Individuals appear to show no decline in fitness and, in stable environments, this type of reproduction is probably beneficial. Treatment with suitable antibiotics results in a reversion to a biparental form producing males and requiring mating.

BACTERIAL DISEASES OF PLANTS

Bacterial diseases of plants are more common in the tropics and subtropics. In the UK we are familiar with the common *Erwinia* soft rots in fruit; *Erwinia amylovora,* which causes fireblight disease; *Xanthomonas* species causing leaf spotting and stem rots; and *Pseudomonas* species responsible for leaf spots and wilts, while *Agrobacterium tumefaciens* causes gall-like structures (Figure 6.39). This latter species has taken on enormous importance in plant breeding techniques because it is used

Figure 6.39   Chrysanthemum stem infected with bacterial crown gall (*Agrobacterium tumefaciens*).

as a molecular biology tool to incorporate specific gene sequences into a plant. Bacterial diseases of plants are sometimes carried by insect vectors, particularly flies, whose larvae are adapted to living in wet, rotting environments.

BACTERIAL DISEASES OF INSECTS

One of the most serious bacterial infections is *Paenibacillus larvae* and *Melissococcus plutonius,* the causative agent of American and European foul brood, respectively—a devastating destroyer of honeybee colonies.

For the last 50 years one of the most important biological control agents has been the bacterium *Bacillus thuringiensis* (*Bt*), used for the control of various pest caterpillars (Figure 6.40), and related strains for control of mosquitoes and some beetles like chafer grubs and Colorado beetle larvae. There are many strains of *Bt* and these cause insect death by paralysis by producing an endotoxin that is released by the bacterium in the gut of suitable hosts. The gene responsible for producing the toxin has been isolated and incorporated into some crop plants, particularly cotton and maize, and makes them resistant to various herbivorous insects. While some people feel uneasy about ingesting insect toxins in their food, others worry that such plants will increase insect resistance to *Bt*. If these genes were to be expressed in floral organs, then nectar and pollen might damage pollinators such as butterflies and moths feeding on insect-pollinated flowers of such GM plants. There is some work that suggests that *Bt* toxins can be released into the soil as the plant breaks down, killing insects such as springtails and thereby slowing the recycling of soil nutrients.

Fungi

Fungal taxonomy has been revised on the basis of DNA and groups such as the oomycetes have been moved out and allied to the algae, while the previous unclassified deuteromycetes are now grouped with the Ascomycota. In general, fungi are considered to have more in common with animals than plants and share some of their biochemistry with the arthropod group, such as their chitin cell walls. While their ancestors arose in Precambrian times (over 1,000 million years ago), most of our modern classes were present by the late carboniferous period (300 million years ago).

Figure 6.40   Caterpillar of large cabbage white butterfly (*Pieris brassicae*) in the latter stages of infection by *Bacillus thuringiensis*.

Figure 6.41   Late blight of potato (*Phytophthora infestans*) showing wilting and necrotic foliage.

The oomycetes (no longer considered true fungi) include a number of well known plant pathogens, such as *Phytophthora* species, responsible for potato blight (Figure 6.41) and root rots; the damping-off disease of seedlings caused by *Pythium* species; and downy mildews, *Peronospora* and related species, found on many crop plants.

The Glomeromycota = Zygomycota are considered primitive fungi comprising simple bread mold species and symbiotic arbuscular mycorrhiza, found in over 80% of all vascular flowering plants. They invade plant roots, increasing the supply of nutrients and plant vigor.

Several *Entomophthora* and related species attack a wide range of insects in temperate climates (Figure 6.42). So far it has proved very difficult to produce these beneficial fungi commercially as biological control agents. This difficulty is caused by the need for these fungi to mate with a different mating strain and produce a resting spore. The mass-production process requires an environment matching that going on inside a living insect body and is difficult to replicate.

Various obligatory endophytic symbiont fungi occur in plants. For instance, *Neotyphodium* spp. (teleomorph *Epichloë*) are seed-borne endophyte symbionts in temperate grasses (Figure 6.43), which

Figure 6.43   Choke disease (*Epichloë typhina*) on wild grass.

can be found distributed throughout the plant and stimulate resistance to herbivores by the induced production of certain alkaloids. Clearly, some endophytes can increase the toxicity of plants to grazing animals and humans, while others may be repellent or beneficial to specialist insect feeders. Some endophyte secondary metabolites like terpenoids may attract parasitoids and predators to damaged plants. There is great interest in this area as inoculation of a crop with a suitable endophyte species, perhaps with its own detrimental genes removed or beneficial ones added, is seen as an alternative and "organic" way to transform the genetic expression of resistance in a crop.

The Ascomycota are a very large group of true fungi comprising yeasts, truffles, and fungal pathogens that cause powdery mildew, apple scab (Figure 6.44), and ergot diseases (see

Figure 6.42   Black currant leaf infested with currant sowthistle aphid (*Hyperomyzus lactucae*) infected with Entomophthora fungus (*Pandora neoaphidis*).

Figure 6.44   Cracked lesions of apple scab (*Venturia inaequalis*) on young apple fruit.

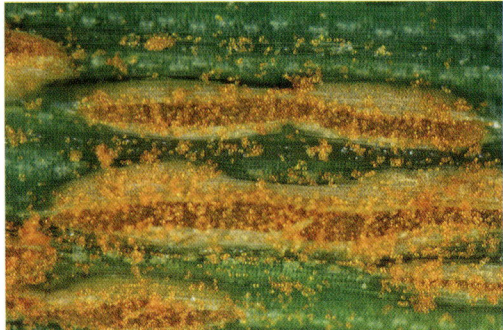

Figure 6.46   Pustules of crown rust (*Puccinia coronata*) on oat leaf.

Figure 6.45   Leaf curling plum aphids (*Brachycaudis helichrysii*) infected with *Lecanicillium lecanii*.

Figure 6.37) as well as being a partner with algae to form most of our familiar lichens.

The Pezizomycotina subgroup contains a number of subtropical species that can be mass produced and used in protected cultivation for control of pests. These include the genera *Lecanicillium* (Figure 6.45), *Beauveria*, and *Metarhizium*, which infect specific insects, and *Paecilomyces* that infect

nematodes. In this subgroup no sexual reproduction has been recorded, so commercial mass production can be undertaken in fermentation tanks similar to those used for yeast production. *Trichoderma* species occur in the soil around plant roots and protect plants by attacking other fungi. They too can be easily cultured and used as a biological control agent against many soil-borne fungal pathogens. In the UK, *Trichoderma* products require registration and must be shown to be of native origin.

The Basidiomycota comprise an equally large group including the familiar mushrooms and toadstools as well as pathogens that cause rusts and smut diseases (Figure 6.46). Many of these fungi show interesting relationships with insects that may help vector them to new host plants. *Piriformospora indica* is a root-colonizing, easily cultured endophyte species that readily increases resistance to root pathogens and induces systemic disease control in many plant hosts. Some see this species as a model for understanding plant resistance mechanisms and others a promising newcomer for disease control, although possibly more suited to warmer tropical climates. It is not currently licensed for use in the UK.

Viruses

Viruses cause some of the most serious and deadly diseases of both plants and animals.

They are not in themselves living organisms, but rather a sequence of genes that can infect living cells and replicate further virus particles. While some animal viruses can spread in water droplets in the air and are inhaled by the host, most plant and animal viruses require an animal vector. We have already mentioned soil nematode vectors, but by far the most common plant virus vectors are found in the sap-feeding Hemiptera group—in particular, the aphids and whiteflies and, to a lesser extent, thrips and plant-feeding mites. The related blood-feeding tick group is one of the most successful virus transmitters to vertebrate animals, as are some fleas, mosquitoes, and biting midges (Figure 6.47).

In some cases the plant viruses are transmitted mechanically from an infected host plant to another on the insect's mouth-parts (Figures 6.48 and 6.49). In others, a replication phase is required, in the insect, before it can reinfect a new plant. As the

Figure 6.48   Bird cherry aphids (*Rhopalosiphum padi*) feeding on barley yellow dwarf virus (BYDV)-infected wheat leaf.

Figure 6.49   Field of wheat showing yellowing foci of infection of BYDV in a wheat crop.

virus-infected plant builds up virus particles inside, it may give off signals that attract more insects to land and probe. This is known to be the case with plants infected by cucumber mosaic virus, where virus-infected plants attract aphids (Figure 6.50). But the virus makes the plants less palatable,

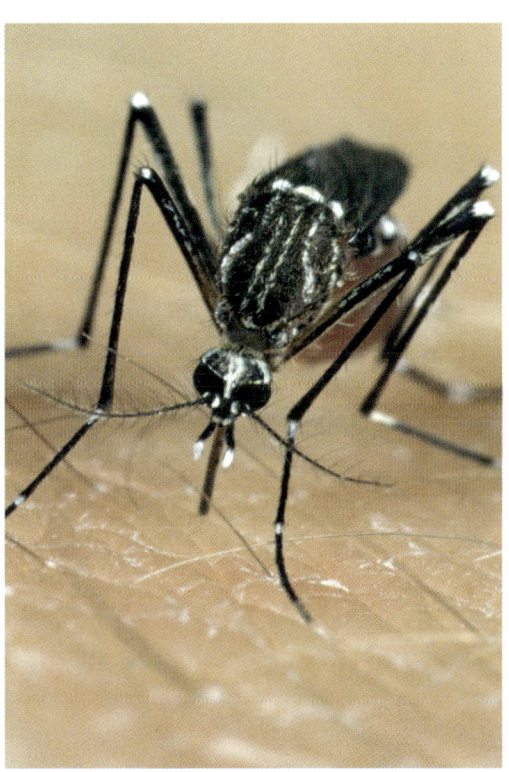

Figure 6.47   Adult Egyptian mosquito (*Aedes aegypti*) feeding on human hand.

MICROORGANISMS

Figure 6.50 Distortion and leaf discoloration of cucumber leaves caused by aphid-transmitted cucumber mosaic virus.

Figure 6.51 Small cabbage white butterfly caterpillar (*Pieris rapae*) infected with granulosis virus (*Pr*Gv).

so the aphids move on. As this is a nonpersistent virus, it is very rapidly spread as a result of this mechanism. Such a mechanism may apply more generally as the nutritional status of a virus-infected plant is impaired, so the insect moves on to infect further plants.

For biological control of insect pests there are only a few commercial viruses available and each is specific to a single insect species. These viruses have to be produced in the laboratory on the appropriate host species, removed from all host material, and then formulated for application as a spray. They work effectively when applied to young larval stages but rapidly break down in sunlight. In the UK a commercial virus is available for control of codling moth (*Cydia pomonella*) on apples and pine sawfly (*Neodiprion sertifer*).

Viruses, like bacteria, must be ingested to infect insect hosts. In sawfly larvae, virus infections are limited to the gut, and disease symptoms are not as obvious as they are in moth caterpillars. In caterpillars, virus particles pass through the insect's gut wall and infect other body tissues. As an infection progresses, the caterpillar's internal organs are liquefied, and its cuticle (body covering) discolors and eventually ruptures. Caterpillars killed by

viral infection appear limp and soggy. They often remain attached to foliage or twigs for several days, releasing virus particles that may be consumed by other larvae (Figure 6.51). The pathogen can be spread throughout an insect population in this way (especially when raindrops help to splash the virus particles to adjacent foliage) and also by infected adult females depositing virus-contaminated eggs.

NATURAL PLANT RESISTANCE MECHANISMS

Whenever we look at a natural, native population of a particular plant in a community, we see a wide range of genetic diversity, sometimes expressed in subtle differences in leaf shape or flower color and differences in growth rate. In a natural habitat we may encounter some individual plants with strong disease symptoms alongside the same species that appears completely resistant.

Today, most crop plant breeders spend their time looking for plants with natural resistance to insects and diseases and breeding new plants with these resistance genes present. However, experience has shown that this resistance, which is often dependent upon a single dominant gene, generally breaks down once the cultivar is grown on a field scale. Where we find no breakdown over many years, it is often because the resistance is polygenic, although very occasionally single genes confer durable resistance. A good example of such resistance is the Tm-2^2 gene widely used in tomatoes, which have remained resistant to tomato mosaic virus for many years.

The epidemiology of plant diseases is largely based on diseases in crops and, in particular, on the role of wind dispersal and water splash, which can lead to catastrophic outbreaks (e.g., the development of potato blight [*Phytophthora infestans*] epidemics). There have been few studies of disease spread in natural and complex plant communities, although there has been more research in crops of mixed cultivars. Wind dispersal may be effective in open crop type habitats but would be less successful in dense and complex natural plant communities.

We know that flowering plants evolved to use insects to move their pollen grains from plant to plant, so it would be reasonable to expect some disease organisms to use pollinators, herbivores, beneficial predators, parasitoids, and ants to be potential vectors as they need to uniquely seek out and move between plants of the same species. Certainly, some recent research has demonstrated that the black ant (*Lasius niger*) can carry fungal spores of *Lecanicillium* to infect aphid colonies that it visits.

Detailed investigation of plant genomes reveals many resistant "R" genes that are able to recognize specific avirulent (Avr) genes in the pathogen and set off a rapid defense reaction. This is an area of active research and is likely to lead to many novel ways of increasing plant resistance (Figure 6.52).

We grow crops for their uniformity, such as a wheat cultivar for its bread-making quality, but a single cultivar crop will sooner or later be overcome with a disease problem. The ideal would be to have a single cultivar that had some genetic diversity for disease resistance. Conventional plant breeding changes too many genes and requires a long period of selection. Using molecular techniques, it is possible to keep the same basic plant cultivar but produce isogenic lines with resistance to different strains of a pathogen. By growing these isogenic lines as a single crop with a mix of resistance genes, we might retain the overall desirable feature of the original cultivar while ensuring that a particular disease outbreak will affect none or only a small proportion of the plants present (Figure 6.53).

We know that it has taken millions of years for specific diseases and herbivores to establish relationships with particular plant families and species. Using molecular techniques to produce a GM crop variety with natural resistance to match the disease organism seems to be an ecologically

Figure 6.52   Research technician working with a wheat experiment in a controlled environment room where light, temperature, and humidity are controlled to monitor plant growth at Rothamsted Experimental Station, Hertfordshire, England.

Figure 6.53   Field researcher assessing experimental crop of rice to determine disease resistance in genetically engineered varieties at the International Rice Research Institute, Luzon, Philippines.

Figure 6.54   Young leaf of fat-hen (*Chenopodium album*) densely covered in protective waxy vesicles.

acceptable practice, as long as it is being used to activate genes that occur naturally in that species. However, some current genetically modified plants use genes like the *Bt* toxin from a completely unrelated taxon that could provide an extremely long-lasting resistance to natural regulating agents. The molecular biologists will have incorporated markers and a mechanism to prevent such plants propagating themselves. To produce a new plant species with no known natural control and the potential to spread like some of our alien imports could be argued as hugely irresponsible and completely unacceptable.

Resistance Mechanisms

Resistance mechanisms fall into three broad categories: antixenosis (repellent), antibiosis (damaging to growth, survival, or reproduction), and tolerance.

Antixenosis is an important defense against insects and other animals typified by physical barriers such as waxy cuticles (Figure 6.54) and sticky, branched, hooked, and dense hairs (trichomes), some of which may break releasing irritant chemicals (Figure 6.55), together with resinous and latex exudates that can inhibit function of mouthparts. Increasing plant resistance to virus vectors can increase the spread of viral diseases within a crop.

Figure 6.55   Dense hairs (trichomes) of stinging nettle (*Urtica dioica*).

Chemical defenses are responsible for antibiosis reactions and can be divided into relatively "permanent" (constitutive) components and induced responses that are only brought into use when stimulated by damage from animals or microorganisms.

Permanent defenses include tannins that inhibit digestion and protein absorption and a wide range of toxic alkaloids that can attack various metabolic pathways, including nerve transmission in animals. Other natural insecticides include azadiractins from neem trees, pyrethrum from chrysanthemum (Figure 6.56), and nicotine from tobacco. Differences in levels of these plant defenses in related plant species cause pests to grow at different rates. Associated predators and parasitoids are smaller and less fecund when developing on stressed hosts. While it

Figure 6.56   Flowering crop of pyrethrum chrysanthemum (*Tanacetum cinerariifolium*) grown to produce pyrethrins—a natural insecticide/acaricide.

would seem sensible to suppose there will be synergism when using plant defenses and biological control together, in practice it seems that this may not be the case.

Cyanogenic glycosides found in many plant families are stored in the cell vacuole and instantly release cyanides when hydrolyzed by beta glucosidase enzyme released by damage to cell cytoplasm. Some insects are able to break down these compounds and some moth species sequester the plant cyanogenic glycosides in their bodies for protection against their attackers.

Similarly, in brassicas, glucosinolates immediately release isothiocyanate (mustard oil) when myrosinase enzyme is released by cell damage. Some cabbage pests sequester the glucosinolates into their own body and use their own myrosinase enzymes to give protection from predators. While the individual caterpillar will be killed, the chemical response teaches the predator to avoid attacking this type of caterpillar again. Insects with these chemical defenses often show yellow and black spots or a stripe warning signal that will be remembered by the predator or, perhaps, is already part of a genetic encoded warning alarm.

Induced defenses often produce an increase of toxic alkaloids of which there are many different types from various plant families. Studies have identified jasmonic acid and sali-

cyclic acid pathways, triggered by wounding, as responsible for producing antidigestion, protease inhibitors against insects and the hypersensitive reactions to pathogens causing localized cell death, respectively. As we have seen, mycorrhiza and specific fungal endophytes play a major role in induced defenses, either by adding to natural plant defenses or being responsible for synthesizing them in the first place. These induced defenses then become part of the system that attracts predators and parasitoids.

Induced responses are related to the cause of the damage and develop accordingly. Thus, damage caused by a caterpillar feeding with its salivary secretions will be recognized as different from that of the adhesion of an egg by an ovipositing insect onto or into the leaf surface. It is these responses that call in the correct and specific parasitoid or predator to deal with the wound-causing herbivore. Such responses can rapidly spread throughout the plant systemically and can then spread to neighboring plants through aerial volatiles and/or perhaps root exudates.

THE FUTURE OF BIOLOGICAL CONTROL

While the control of plant pests will still involve predators and parasitoids, it is clear that plant protection chemicals will be used to monitor populations, change volatile profiles, and simulate kairomones and pheromones to attract and repel. A new science of molecular ecology can orchestrate the calling signals by switching genes that are responsible for plant defenses and beneficial attraction on or off. GM techniques can replace genes lost in the breeding process and produce a basic single cultivar with an enhanced genetic diversity for disease resistance.

The present overuse of crop fungicides needs to be reevaluated as they may destroy the microorganisms that can be responsible for the plant's chemical defenses and attractants.

There is much more to do with mass production and encouraging the role of microorganisms associated with roots such as various mycorrhizal associations and understanding and manipulating the role played by plant endophytes.

Producing crops with companion plants and ground cover is essential to providing alternative food and shelter for beneficials.

Replacing conventional plowing with direct drilling enables the carryover of beneficials, supporting plants and microorganisms from one season to another and preserving the integrity of the soil.

Future agricultural research will need to focus on plant, insect, and disease community structure and the signaling language that drives population stability.

Glossary

abdomen: rear part of body of an arthropod

alate: winged

antagonist: enemy

antenna/antennae: sensory organ on insect's head; "feeler"

apodous: insect larvae without legs

aposematic: warning (colors)

apterae: wingless

arrestment: function of host/prey kairomone to retain a beneficial in the close vicinity of its host/prey

arthropod: animal with external skeleton, jointed legs, and segmented body

banker plant system: system where beneficial insects are encouraged onto plants that are then moved to within the main crop to boost biological control numbers

bioindicator: species used to detect changes

blastospore: asexual fungal resting spores produced by budding

carapace: hard covering of body

cephalothorax: forward section of arachnid body whereby the head and thorax are joined

chaetotaxy: arrangement of bristles on an organism's body

cleptoparasite: insect that feeds its young on food collected by another

cocoon: silky case spun by insect larva for protection of pupa

colonist: species attracted into an area by prey

conidiophore: fungal fruiting body

conidiospore: asexually produced fungal spore produced on a conidiophore

conidium: fungal spore

conventional: referring to the common use of pesticides and growing techniques as opposed to organic methods in which only naturally derived pesticides and methods are permitted

crawler: first-stage larva of usually sedentary insect

cuneus: triangular or wedge-shaped region at the end of the forewing of some Hemiptera

cuticle: waxy layer covering the body

damping off: range of fungal disease of plant roots

dauer larvae: nematode larvae in stasis for adverse conditions; also infective larval stage of parasitic nematodes

detritivore: animal that eats waste

deutonymph: second nymphal stage of mites, after egg and protonymph

diapause: hibernation stage induced by low temperatures and shortening day length during autumn, whereby the organism overwinters

dicotyledon: flowering plant that has an embryo with two cotyledons (embryonic leaves)

diurnal: active during the day

ecosystem: biological community of interacting organisms and their environment

ectoparasitoid: parasitoid that develops outside the body of the prey

elytron/elytra: wing case

endoparasitoid: parasitoid that develops within the body of the prey

entomopathogenic: causing disease in insects

epidermis: outer layer (of skin, leaf)

epigyne: female spider genital area

epizootic: (disease) widespread throughout a population

estivation: summer dormancy (hibernation)

facultative parasitoid: parasitoid that can complete its life cycle independently of a host

filiform: thread-like

frass: fine fecal pellets produced by phytophagous insects and mites

fungivore: animal that eats fungi

gall: deformed growth on a plant produced by insects, fungi, mites, or nematodes

Gram-positive bacteria: group of bacteria that retain crystal violet stain identifying them as having a cell wall in a thick layer of peptidoglycan

gregarious parasitoid: several eggs or larvae developing from a single host

haltere: club-shaped fly balance organs, modified hindwings

haemocoel: cavity within arthropod body holding major organs

haemolymph: fluid in an insect serving as its blood

hermaphrodite: containing both sets of sexual reproductive organs

honeydew: sugary waste of many phloem-feeding insects

hydroponics: process of growing plants without soil

hyperparasitoid: parasitoid that develops in or on another parasitoid

idiobiont: parasitoid that paralyzes its host and develops (usually) next to the host as an ectoparasitoid

instar: stage of life between any two molts, adult being the final instar

integrated crop management (ICM): integration of chemical, biological, cultural, and physical control methods to reduce input for sustainable crop production

integrated pest management (IPM): use of commercially raised beneficial insects with compatible pesticides

intraguild predation: predators sharing the same prey species that may attack and kill each other

invertebrate: animal lacking a backbone

kairomone: scent produced by a host organism that attracts a prey organism

koinobiont: parasitoid that develops inside its host as the host continues to develop, killing the host when the parasitoid reaches maturity; see also idiobiont

lamina: thin layer (of leaf)

larva/larvae: immature insect following emergence from the egg; principally insects having a complete metamorphosis and before the pupal stage

mandible: jaw or mouthpart of an arthropod; it may be toothed for biting or needle-shaped for sucking

maxilla: part of arthropod mouthpart found just behind the jaws, used for handling and sucking chewed food

mineral wool: synthetic growing medium made from reconstituted stone fibers formed into blocks with good air- and water-holding capacity for growing plants; also known as stone wool

molt: shedding of outer skin

multivoltine: having several broods in one season

mycoinsecticide: biological insecticide based on fungal metabolites

neonate: newly hatched or born

nocturnal: active at night

nymph: immature form of insect or mite that closely resembles the adult; no pupation occurs to become an adult

ocularium: tubercle bearing a pair of sideways-facing simple eyes

opercula: flap or lid covering aperture

oviposition: process of laying an egg

ovipositor: tubular egg-laying organ of insect

Palearctic: biogeographic zone including Europe, parts of North Africa, and most of Asia

palp: segmented sense organ at the mouth of an arthropod

parasitoid: organism that lives in or on another organism for part of its life cycle and eventually kills the host

parthenogenesis: asexual production of young without fertilization of eggs

peptidoglygcan: cross-linked complex of polysaccharides and peptides in the outer cell walls of bacteria

perennial: plant that lasts through several growing seasons

petiole: ring-like second abdominal segment of a wasp

pheromone: chemical sex attractant used by males

phloem: nutrient-conducting tissue of plants

phytophagous: plant eating

pneumostome: pore at rear of slug mantle leading to "lung"

polyembryony: multiple individuals developing from an egg polyphagous: eating a variety of organisms

proboscis: elongated, sucking mouthparts of an insect

pronotum: main part of dorsal surface of first thoracic segment

proteolytic enzyme: breaks down chains of protein to smaller units

protonymph: first stage after mite egg hatch

pupa/pupae: inactive immature stage of an insect in which the larva metamorphoses to the adult

puparium: barrel-shaped case formed from last larval skin in which (mainly fly) pupa is hidden

pyrethrum: naturally occurring pesticides from species of chrysanthemum, usually with short persistence

refugium/refugia: area in which a population of organisms can survive during unfavorable conditions

resident: species present throughout the growing season

rhizoid: like a root

rostrum: beak-like part of insect or arachnid mouthparts

seasonal inoculative biological control: release of natural enemies in an area in which they are expected to establish during the main period of pest activity

senescence: growing old

seta: stiff hair or bristle

spinneret: spinning organ of an arachnid

stone wool: synthetic growing medium made from reconstituted stone fibers formed into blocks with good air- and water-holding capacity for growing plants; also known as mineral wool

stylet: piercing mouthpart of an insect and stinger of wasp

sympatric distribution: closely related (sister species); found in the same locality

synthetic pyrethroid: man-made insecticides based on natural pyrethrum but with broad-spectrum activity and long persistence

tarsus/tarsi: foot or fifth joint of leg of an insect

temperate: of mild climatic conditions

thorax: middle part of the body of an arthropod to which legs and wings (if present) are attached

top fruit: fruit grown on trees

transconjugation: joining of two genetic strains

trichome: sticky hair of plant

understory: layer of vegetation under main canopy of trees

univoltine: having only one generation each year

vena spuria: false vein

virion: single virus particle

viviparous: producing live young

References

Aldridge, C. and Carter, N. 1992. The principles of risk assessment for non-target arthropods: A UK registration perspective. Interpretation of pesticide effects on beneficial arthropods. *Aspects of Applied Biology* 31: 149–156.

Baur, R., Remund, U., Kauer, S., and Boller, E. 1998. Seasonal and spatial dynamics of *Empoasca vitis* and its egg parasitoids in vineyards in Northern Switzerland. Proceedings of the IOBC/WPRS Working Group Viticulture, March 4–7, 1997, Godollo, Hungary. *IOBC/WPRS Bulletin* 21 (2): 71–72.

Boller, E. F. 2001. Functional biodiversity and agro-ecosystems management: 1. Identified information gaps. Integrated fruit production. *IOBC/WPRS Bulletin* 24 (5): 1–4.

Bravenboer, L. and Dosse, G. 1962. *Phytoseiulus riegeli* Dosse als Predator einiger Schadmilben aus der *Tetranychus urticae* gruppe. *Entomologia Exp. Appl.* 5: 219–304.

Brown, M. W. 2001. Functional biodiversity and agro-ecosystems management: 2. Role in integrated fruit production. *IOBC/WPRS Bulletin* 24 (5): 5–11.

Brown, R. A. 1989. Pesticides and non-target terrestrial invertebrates: An industrial approach. In *Pesticides and non-target organisms*, ed. P. C. Jepson, 19–42. Wimbourne: Intercept.

Chandler, D. 2008. The consequences of the "cut off" criteria on pesticides: Alternative methods of cultivation. Policy Department Structural and Cohesion Policies, Agriculture and Rural Development. European Union, Brussels.

Cilgi, T. and Vickerman, P. 1994. Selecting arthropod "indicator species" for environmental impact assessment of pesticides in field studies. *Aspects of Applied Biology* 37: 131–140.

Cross, J. V., ed. 2002. Guidelines for integrated production of pome fruits in Europe. Technical guideline III. *IOBC/WPRS Bulletin* 25 (8), 45 pp.

Cross, J.V. and Hall, D.R. 2009. Exploitation of the sex pheromone of apple leaf midge *Dasineura mali* Kieffer (Diptera: Cecidomyiidae) Part 1.

Development of lure and trap. *Crop Protection* 28(2): 139–144.

Greig-Smith, P. 1991. The Boxworth experience: Effects of pesticides on the fauna and flora of cereal fields. In *The ecology of temperate cereal fields*, eds. L. G. Firbank, N. Carter, J. F. Darbyshire, and G. R. Potts. London: Blackwell Scientific.

Hussey, N. W. 1985. History of biological control in protected culture. In *Biological pest control. The glasshouse experience*, eds. N. W. Hussey and N. Scopes, 11–22. Poole, UK: Blandford Press.

Hussey, N. W., Read, W. H., and Hesling, J. J. 1969. *Pest control: Materials and methods. The pests of protected cultivation*, 9–43. London: Edward Arnold (Publishers) Ltd.

Luczak, J. 1975. Spider communities of crop fields. *Polish Ecological Studies* 1: 93–110.

Nyffeler, M. and Benz, G. 1987. Spiders in natural pest control: A review. *Journal of Applied Entomology* 103: 321–339.

Nyffeler, M., Sterling, W. L., and Dean, D. A. 1994. Insectivorous activities of spiders in United States field crops. *Journal of Applied Entomology* 118: 113–128.

Paoletti, M. G. and Bressan, M. 1996. Soil invertebrates as bioindicators of human disturbance. *Critical Reviews in Plant Sciences* 15: 21–62.

Piggott, S. J., Clayton, J., Gwynn, R., Matthews, G.A., Sampson, C., and Wright, D. J. 2000. Improving folia application technologies for entomopathogenic nematodes. Workshop proceedings; University of Ireland, May, 13–15, 2000, pp. 119–127.

Samu, F., Tóth, F., Szinetár, C., Vörös, G., and Botos, E. 2001. Results of a nation-wide survey of spider assemblages in Hungarian cereal fields: Integrated control in cereal crops. *IOBC/WPRS Bulletin* 24 (6): 119–127.

Shaw, R. H., Bryner, S., and Tanner, R. 2009. The life history and host range of the Japanese knotweed psyllid, *Aphalara itadori* Shinji: Potentially the first classical biological weed control agent for the European Union. *Biological Control* 49: 105–113.

Sunderland, K. D., Chambers, R. J., Helyer, N. L., and Sopp, P. I. 1992. Integrated pest management of greenhouse crops in Northern Europe. *Horticulture Reviews* 13: 1–47.

Sunderland, K. D., Fraser, A. M. and Dixon, A. F. G. 1986. Distribution of linyphiid spiders in relation to capture of prey in cereal fields. *Pedobiologica* 29: 367–375.

Toft, S. 1989. Aspects of ground-living spider fauna of two barley fields in Denmark: Species richness and phenological synchronisation. *Entomologiske Meddelelser-Entomologisk Forening Kobenavn* 57: 157–168.

Topping, C. J. and Sunderland, K. D. 1992. Limitation to the use of pitfall traps in ecological studies exemplified by a study of spiders in a field of winter wheat. *Journal of Applied Ecology* 29: 485–491.

van Gestel, C. A. M. and van Brummelen, T. C. 1996. Incorporation of the biomarker concept in ecotoxicology calls for a redefinition of terms. *Ecotoxicology* 5: 217–225.

van Lenteren, J. C. (2011). The state of commercial augmentative biological control: Plenty of natural enemies, but a frustrating lack of uptake. Published with open access at Springerlink.com. BioControl DOI 10.1007/s10526-011-9395-1

Vickerman, G. P. 1992. The effects of different pesticide regimes on the invertebrate fauna of winter wheat. In *Pesticides and the environment: The Boxworth study*, eds. P. Grieg-Smith, G. A. Frampton, and A. Hardy. London: HMSO.

Further reading

Alford, D. V. 2007. *A color handbook; pests of fruit crops*. London: Manson Publishing. ISBN 9781840760514.

———. 2012. *A color handbook; pests of ornamental trees, shrubs and flowers*. London: Manson Publishing. ISBN 9781840761627.

Bellows, T. S. and Fisher, T. W. 1999. *Handbook of biological control*. San Diego, CA: Academic Press.

Biddle, B., McKeowin, A., and Cattlin, N. 2007. *A color handbook; pests, diseases and disorders of peas and beans*. London: Manson Publishing. ISBN 9781840760187.

Boller, E. F., Vogt, H., Ternes, P., and Malavolta, C. 2005. Working document on selectivity of pesticides. IOBC/WPRS internal newsletter no. 40.

Bridge, J. and Starr, J. 2007. *A color handbook; plant nematodes of agricultural importance*. London: Manson Publishing. ISBN 9781840760637.

Cross, J. V., Innocenzi, P., and Hall, D. R. 2000. Integrated production of soft fruits. *IOBC/WPRS Bulletin* 23 (11): 67–72.

Dreistadt, S. H. 1994. Pests of landscape trees and shrubs: An integrated pest management guide. State-wide IPM Project, University of California, Division of Agriculture and Natural Resources. Publication 3359, Oakland, California.

Ellis, P. R., Entwistle, A. R., and Walkey, D. G. A. 1993. *Pests and diseases of alpine plants*. Pershore, Worcestershire, UK: Alpine Garden Society.

Flint, M. L. 1990. Pests of the garden and small farm: A growers' guide to using less pesticide. State-wide IPM Project, University of California, Division of Agriculture and Natural Resources. Publication 3332, Oakland, California.

Gill, S., Clement, D. L., and Dutky, E. 1999. *Pests and diseases of herbaceous perennials: The biological approach*. Batavia, IL: Ball Publishing.

Hoffman, M. P. and Frodsham, A. C. 1993. *Natural enemies of vegetable insect pests*. Ithaca, NY: Cornell University.

Hoy, M. A., Cunningham, G. L., and Knutson, L., eds. 1987. *Biological control of pests by mites*. Berkeley: University of California.

Hussey, N. W. and Scopes, N. E. A. 1985. *Biological pest control: The glasshouse experience*. Poole, Dorset, UK: Blandford Press.

Jervis, M. and Kidd, N. 1996. *Insect natural enemies, practical approaches to their study and evaluation*. London: Chapman & Hall, 491 pp.

Koike, S. T. Gladders, P., and Paulus, A. O. 2006. *A color handbook; vegetable diseases*. London: Manson Publishing. ISBN 9781840760750.

MacGill, E. I. 1934. On the biology of *Anagrus atomus*, an egg parasitoid of the leaf-hopper *Erythroneura pallidifrons*. *Parasitology* 26: 57–63.

Mahr, D. L. and Ridgeway, N. M. 1993. *Biological control of insects and mites, an introduction to beneficial natural enemies and their use in pest management*. Madison: University of Wisconsin.

Malais, M. H. and Ravensberg, W. J. 2003. *Knowing and recognizing: The biology of glasshouse pests and their natural enemies*. Rodenrijs, the Netherlands: Koppert, B.V.

Nechols, J. R., Andres, L. A., Beardsley, J. W., Goeden, R-D., and Jackson, C. G. 1995. Biological control in the western United States: Accomplishments and benefits of regional project W-84, 1964–1989. University of California, Division of Agriculture and Natural Resources. Publication 3361, Oakland, California.

Olkowski, W., Daar, S., and Olkowski, H. 1991. *Common sense pest control: Least toxic solutions for your home, garden, pets and community*. Newtown, CT: Taunton Press.

Powell, C. C. and Lindquist, R. K. 1997. *Ball pest and disease manual: Disease, insect and mite control on flower and foliage crops*. Batavia, IL: Ball Publishing.

Raupp, M. J., Van Driesche, R. G., and Davidson, J. A. 1993. *Biological control of insect and mite pests of woody landscape plants: Concepts, agents and methods*. Baltimore: University of Maryland.

Samuels, G. J., Ismaiel, A., Bon, M. C., De Respinis, S., and Petrini, O. 2010. *Trichoderma asperellum sensu lato* consists of two cryptic species. *Mycologia* 102: 944–966.

Shipp, L., Kapongo, J-P., Kevan, P., Sutton, J., and Broadbent, B. 2008. Using bees to disseminate multiple fungal agents for insect pest control and plant disease suppression in greenhouse vegetables. *IOBC/WPRS Bulletin* 32: 201–204.

Steiner, M. Y. and Elliott, D. P. 1987. *Biological pest management for interior plantscapes*, 2nd ed.

Vegreville, Alberta, Canada: Alberta Environmental Center.

Wale, S. Platt, B., and Cattlin, N. 2008. *A color handbook; diseases, pests and disorders of potatoes*. London: Manson Publishing. ISBN 9781840760217.

Watson, A. K., ed. 1993. *Biological control of weeds handbook*. Champaign, IL: Weed Science Society of America.

Yepson, R. B., Jr., ed. 1984. *The encyclopedia of natural insect and disease control*. Emmaus, PA: Rodale Press.

Useful websites

The following are some useful World Wide Web sites (several contain hyperlinks to each other as well as further useful sites).

Agricultural Research Services, US Department of Agriculture: http://www.ars.usda.gov

American Phytopathological Society http://www.apsnet.org/online/archive.asp

Animal and Plant Health Inspection Service; hotlinks to other biological control pages: http://www.aphis.usda.gov/nbci/hotlinks.html

Auburn's Biological Control Institute (BCI), Auburn University: http://www.ag.auburn.edu/bci/bci.html

Biological control at the University of California at Riverside, CA: http://www.biocontrol.ucr.edu/

Commercial biological control producers and suppliers World Wide listings: http://www.agro-biologicals.com/index.html

Commercial biological control producer and supplier: http://www.bcpcertis.com

Commercial biological control producer and supplier: http://www.biobest.be

Commercial biological control producer and supplier: http://www.entocare.nl/engels

Commercial biological control producer and supplier: http://www.koppert.nl/english

Commercial biological control producer and supplier: http://www.SyngentaBioline.com

Commercial specialist producer and supplier: http://www.wyebugs.co.uk

Florida Agricultural Information Retrieval System, University of Florida: http://hammock.ifas.ufl.edu/

Frank Koehler: http//www.koleopterologie.de/gallery (Arbeitsgemeinschaft Rheinischer Koleopterologen)

Glossary of expressions in biological control: http://University of Florida.IPM-143IN673

IPM experts directory: http://www.ipmalmanac.com/experts/index.asu

International Biocontrol Manufacturers Association: http://www.ibma-global.org

International Organization for Biological Control: http://wwwiobc.global

Midwest biological control news: http://www.entomology.wisc.edu/mbcn/land609.html

Natural History Museum, UK: http://www.nhm.ac.uk

New York State Agriculture Experimental Station; a guide to the natural enemies of North America: http://www.nysaes.cornell.edu/ent/biocontrol

North Carolina's National IPM Network by the North Carolina State University: http://ipm-www.ncsu.edu

Ohio Florists Association, United States: http://www.ofa.org

Royal Botanic Gardens, Kew, UK: http://www.rbgkew.org.uk

Royal Horticultural Society, UK: http://www.rhs.org.uk

United States Department of Agriculture: http://www.usda.gov

University of California IPM home page, University of California at Davis: http://www.ipm.ucdavis.edu

University of Nebraska Cooperative Extension: Biological control of insect and mite pests: http://ianrwww.unl.edu/ianr/pubs/extnpubs/insects

University of Purdue Cooperative Extension: Purdue's Biological Control Laboratory: http://www.entm.purdue.edu/entomology/bclab/biocontrol.html

Index